John Oliver Bellville

Thorns and Roses

John Oliver Bellville

Thorns and Roses

ISBN/EAN: 9783337419974

Printed in Europe, USA, Canada, Australia, Japan

Cover: Foto ©berggeist007 / pixelio.de

More available books at **www.hansebooks.com**

THORNS AND ROSES

BY

JOHN OLIVER BELLVILLE

𝕰𝖛𝖆𝖓𝖘𝖛𝖎𝖑𝖑𝖊
KELLER PRINTING CO., PUBLISHERS
1895

COPYRIGHT 1895
BY
JOHN OLIVER BELLVILLE

TO FRANK M. GILBERT

No, sever not sweet friendship's chain;
Here are my deeds, my words are vain.
For he as e'er a true friend should
Condoned my faults and praised my good.

CONTENTS.

	PAGE
THORNS AND ROSES	1
TERRAE FILIUS	5
DESPAIR	6
VINDICTIVENESS	7
THE PASSING OF THE SOUL	8
NOVEMBER	13
STARVATION	14
CONSCIENCE	15
THEN WOULD I ONLY KNOW	16
THE PANACEA	19
THE MUSE	25
A GRIEF AND A BLESSING	28
SPIRIT OF DEVOTION	29
A PLEA FOR FRIENDSHIP	31
THE SEPARATION	32
SONGS OF CHILDHOOD	34
MEDITATION	36
DECEPTION	38
CALL THEM BACK	39
TRUTH AND ERROR	41
EDGAR WELLTON COOLEY	43
IS THERE A SWEETER STRAIN?	45
BURNING OF THE STEAMER WELLINGHAM	46
THE QUARTER THAT HE SAVES	49
IN THE DARK	50
MAN	52
THE VISIBLE PANDEMONIUM	56
WHERE THE BATTLE WAS FOUGHT	60
THE HEAD LINES OF A LETTER	61
A LESSON IN MYTHOLOGY	63
CUSTER'S LAST BATTLE	65
WHEN IT SNOWS	68
THE DREAMERS	70

	PAGE
The War	71
In a World Born to Doubt	72
'Tis Sweet	74
A Song of a Joy Returned	76
Take Life as It Is	77
Slander	80
Upward	81
Jealousy	82
In Slumber Land	84
The Banquet	86
"When Lovely Woman"	90
A Strange Fair Lady	92
A Woman's Worth	94
Spring	98
How Sweet to be Folded Away	100
Constancy	103
A Valentine	104
How Sad It Was	105
A Last Good Bye	107
It is Said Yet Not True	109
When the Last Turbid Tide	111
Is Kissing an Evil?	112
Two Vessels	114
Oh, Where is the Torch?	116
Oh, Take My Heart	118
When the Heart's Last Emotion	120
A Summer Evening	121
Do I Love You?	123
As We Strolled in Our Sleep	125
On the World	127
The Voice of Labor	141
An Appeal to Patriotism	146
The Wanderer	150
In Chains	155
Labor Day	160

THORNS AND ROSES.

A magical artist, with masterly hand,
  Painted a picture of light and of shade—
Light as the water, shade as the land—
And Oh, how rich, sublime and grand,
  That beautiful picture the artist made!

But it had not tints to enchant the view,
  Or to liven its shade or mellow its light
So the artist searched his soul, and drew
  From the universe, each pigment mite,
And skillfully blended them with dew.

And when they were blended to please his eye,
   And serve his especial use and need,
He breathed upon them with a sigh,
   Transforming and turning them into seed
With a life and soul to never die.

Then, turning to where the picture hung,
   He gazed upon its light and shade,
And praised its grandeur, as it swung,
   And blessed the work that he had made,
As the stars and sun together sung.

And as the music rolled through the world,
   Praising his magic and splendid worth,
He scattered the seed, by light impearled,
   With hand divine, and a sense of mirth,
This artist God, o'er the picture—Earth.

Then he breathed on it and sighed again,
    And it bloomed in beauty, life and grace;
And its tints He enameled with mist and rain,
    As He hung it out in the world's wide space,
Linked to the stars with a mystical chain.

The thorn sprang first, and then the rose
    Bloomed from the seed His hand had sown,
But His counselors swore, as did His foes,
    That the work had best been left alone,
In its light and shadow of repose.

And the furies blew on it their breath,
    And lashed it with their ebon wings,
And the rose-bloom swooned in the lap of death,
    But they could not waste the thorn that stings.
Then Satan scowled and gnashed His teeth,

And said: "I will make a battle ground
  Of this scene, and people it with kings,
And scatter o'er it, and around,
  All manner of flying and creeping things,
And good and evil shall abound."

Smiles, Comfort, Gladness, Love, and Light,
  Shall represent each rose that's born;
Hate, Tears, and Sadness, Scorn and Night,
  Shall be as emblems of the thorn,
  O'er which the human race shall mourn!

TERRAE FILIUS.

As a drop of rain in the mighty deep;
　As a leaf blown from a tree;
As a breath of mist in night's long sleep;
　As a grain of sand of the sea;
As a spark of light from eternal fires;
　A ray from the heart of the sun;
As a broken note from eternal lyres,
　Is man, when his race is run.

As a breath of odors of every flower;
　As a shadow from night's great scroll;
As a second from the eternal hour,
　Is the span of a human soul.
He's fair as a lily to human sight
　But alas! a smile—a nod,
And then as a star that has spilled it's light
　He sleeps in the lap of God.

DESPAIR.

Then came Despair to one who sadly sat
   Amid her tears and moaned for having sold
   The lily of her virtue for the trash called gold.
It knelt beside her, while grim hunger lingered at
   Her window, gazing at her with a deathless stare,
   So weird and hollow-eyed that e'en Despair
   Sobbed piteously and wept, and left her there.

And winged its way to the most lofty peak
   Of the dull mountains wrapped in ice and snow,
   And there, amid the darkness and its woe,
And the dim stars, and north-winds, cold and bleak,
   It raved as raves a lost soul cast away,
   And prayed the sun, approaching near to day,
   To hide his face, and never cast a ray
Upon an earth where mortals are so weak.

VINDICTIVENESS.

In my soul's treasure-store thou wast its richest gem ;
   I wore thee as a charm about my heart ;
   Thy life I felt, was of my own a part,
But now I cast thee from me as a limb
   Cast from a tree within a mighty storm,
   And hate most bitterly thy haunting face and form,
Because of thy despised seductive art.

Forgive thee? No! Go swart the lily's face
   Or crush the fragrance from a red rose-bud
   Beneath thy foot—press out its living blood ;
Then canst thou place it back within the vase
   From whence it came as perfect and as fair
   And fragrant as when first thou foundst it there?
No! Nor canst thou its lovely tintings trace.

THE PASSING OF THE SOUL.

In the throes of deathless dreaming;
In the dismal darkness streaming;
In deep chaos with no gleaming
    Ray of light to guide my mind,
Groped I, hoped I; knelt imploring
For one ray of light sent soaring
To this ghoulish, rayless region
    Where I dwelt in terror blind.

Then a spirit, sad and dismal,
From Avernus depths abysmal,
Came with ebon urn, baptismal,
    And made darkness deeper grow;
Gibbed at me and dragged me, speechless,
Downward to the depths so reachless,
That the tireless wings of fancy
    To their lowest could not go.

Darker, darker, deeper, deeper,
"Who, O God," I shrieked, "is keeper
Of the midnight where no sleeper
   Closes eye-lids on its shores?"
"Who?" Eumenides made answer.
"I am he, with sword and lance sir,
I exclude the light forever
   With my bronzed, Plutonian doors."

"You?" I cried; then felt a chilling—
Something o'er my soul came, thrilling
It as tho' the storm-gods, willing,
   Spread their lightnings over me.
Then the darkness hovered deeper,
And I cursed each swarted keeper—
Cursed, then prayed for them to lead me,
   Into light that I might see.

All in vain! Incarcerated
As a Judas, scourged and hated,
With Azazel, relegated,
   I was, as Prometheus, bound—
Chained to hunger's lorn rock, freezing,
Hopeless since there's no appeasing
Of the gaunt and ghostly vulture
   Feasting on my liver's wound.

"O, Eumenides, appalling,
 Are these chains about me, galling;
 And their weight around me falling
   Filches from me freedom's breath.
 Still they eat," I said, "and clanker,
 Gnawing as an ichoris canker,
 Gnaws and claws to reach the vital,
   Like the serpent-tongue of death!"

"Strike them, let me feel their breaking!
 Strike them, let me see their shaking!
 Let me feel no more the aching
   Of the clotting of life's wine!
 I implore thee grant my fleeing
 From the fire-crypts of my being—
 From this petrous pile of torture,
   Unto Calvary's rock divine."

"And I tell thee, prince Satanic,
 Tho' thou hast a strength Titanic,
 Thou shalt see, as by Volcanic
   Greatness, these gyves fall from me!
 Tho' thy hideous witches flout me—
 Serpents hiss and coil about me,
 I will trample them and crush them,
   And forever more be free!"

"From Alaraf's awed disaster;
From the vengence of Alaster,
Take me, if there's God or Master!
    O'er the rainbow bridge of joy.
Bear me, as was born Anchises
On the shoulders of Æneas
For my soul burns as the burning
    At the awful siege of Troy!"

Then a voice spake from the rayless
Shores of this weird starless, dayless,
Torturing region of Abayless—
    Spake as melody to me:
"Ye who grope for God, O mortal,
Now stand at His golden portal
But, save by the orbs of spirit,
    Thou canst not His wonders see."

"Darkness has no power, eternal,
And no sombre fiend, infernal,
Can dispel the light, supernal,
    Flowing from the mind and soul.
God is in you—He's the yearning
Of your inner soul—the burning
Of the cerements that bind it
    As it grasps a higher goal."

Then in mists of mystic splendor,
Came a spirit, meek and tender,
And the heavens seemed to lend her
    All the sweetness of a dove;
And my spirit, at her wooing,
Left its clay and went pursuing—
Went and left the dismal darkness
    For the light of Truth and Love.

Passed Amaimons' power restraining;
Passed Bellerus' bonds retaining;
Passed Asmodeus, profaning,
    O'er the mad maelstroms of fate.
Passed I to the Asgard's glory,
Glorious as the rainbow story,
O'er the pearl and gold of Bifrost,
    And the spans of Alsirat.

There I learned there is no evil;
All is good, if we believe all;
All are forces, not of Devil
    But one centralizing good;
And that God is only real—
In the highest sense ideal—
Not in being, but in spirit,
    And as such is understood.

NOVEMBER.

Athwart the wastes of dull November sky,
  Amid the clouds that hurriedly move on
  Like snow-flakes scurrying over heaven's lawn,
The wild-goose trails his way and passes by,
Combing the wings of night with plaintive cry,
  And pulling the snow-storm with him at the dawn,
  To chill each flower's soul when he is gone
To that far clime where north-winds rarely sigh.

And as the day unfolds her golden doors,
  Jenco Hyemalis in their dark disguise
  Drift in like feathered waifs from Paradise,
Spilling their broken notes upon earth's shores
Where, stooping low, the noiseless snow-queen pours
  White flakes, like frozen tears from angel's eyes.

STARVATION.

They sat together in a dismal cell
  Did he and Hunger; mutely sat and gazed
  Upon each other until he grew dazed.
Then weakness coiled about him and he fell,
In the deep darkness of that awful hell,
  And naught but his poor shrunken hands were raised,
To guard his dying lips from such a spell.

And bit, by bit, he gnawed the flesh from these
  Until their naked bones like ivory swung
  About his head; then grim, gaunt hunger wrung
The fleshless joints it's craving to appease.
And he, poor wretch, felt his heart's pulses freeze,
  But murmured not, for Death had chilled his tongue.

CONSCIENCE.

Upon a rose-hued table in a room
   Lay an open book, so spotless white and fair
   I knew that God's own hand had placed it there
And left it stainless as the lily's bloom.
At length poor Love in melancholy gloom
   Chanced by. He entered in to offer prayer
   And wrote upon the first leaf, unaware
That he had soiled the page and sealed his doom.

But when he came again, Deceit and Sin,
   And Lust and Avarice and Appetite
   Came also with him, and with him did write
Their strange cantations on the leaves therein.
And lo, that book of Conscience, that had been
   As white as snow, was now as black as night.

THEN WOULD I ONLY KNOW.

Out from the veils of darkness,
   Where ignorance broods and pines
And the ghouls of superstition
   Are impaled by false divines,
I climbed away from the shadows
   To the sun-swept plains of truth,
And gazed back on the chaos
   And confusion of my youth.

While the great arch-light of the future
   Beckoned my spirit on
To a universe more splendid
   Than the dreams of a golden dawn—
Where man was at rest and happy,
   Not groping in dread of a curse,
But his soul's harp joined in the music
   Of the lays of the universe.

There were no creeds before me—
  I'd trampled them into the mire;
And dead theology I'd burned
  On cruelty's funeral pyre,
For I shuddered at a religion
  That begets a selfish greed;
That stains the earth a crimson
  And leaves poor man in need!

Yet God, O! God, who art Thou?
  That man shouldst personize
And give Thee human attributes
  And ever-seeing eyes?
Who? Who? I ask the sunbeams;
  I ask the stars of night;
I ask the granite peaks that rise
  Beneath the orbs of light;

The flowers and their fragrance;
  The happy birds that sing,
And each one answers, to my soul:
  "God dwells in everything."
Yet man has never known Him,
  Because He dwells too near
And Superstition ever points
  To a distant unknown sphere.

Oh, had I erudition
   To analyze all thought,
And trace time's evolution
   Till it ended in a blot;
Know every passion of the heart,
   Of sea, of air, of sod—
Then, and only then I'd know
   And understand a God.

THE PANACEA.

Was she mystic? Was she chthonic?
True, she bore a love, platonic,
But she woo'd me like a fairy
   Till I worshiped at her feet;
Then with eyes upon me beaming,
From which softened glows were streaming,
She proved that she was wiser
   Than the sagest thesmothete.

"Taste," said she, "this is the liquor;
This the sweet, ethereal ichor
That the gods imbibed with blessings
   At the fount of Hippocrene."
Then I took the cup and tipped it—
Tipped it to my lips and sipped it
And entranced I saw such visions
   As no mortal eyes had seen.

Down a secret passage, winding
Through a labyrinth, all blinding
With radiant diamond mirrors
   That reflected all her charms,
She lead me, and I listened,
As the lights about me glistened,
Till, as Proteus I saw her,
   In at least an hundred forms.

"Speak not," said she, "I am Ariel,
I assume all shapes, ideal,
And the soul that follows my soul
   Through this labyrinth, must see
Every wonder that is mystic,
Till his sense idealistic
Is lost in the Shechinah
   Of man's immortality."

Here's the cave of Canacea
Leading to the Panacea,
Enter, kiss her ring and listen
   To the music of her words,
And forever, by commanding,
You will have an understanding
Of all races and all kindreds,
   And the language of the birds."

I obeyed as I was bidden,
And the mystery, long hidden
In the wasted tower of Babel,
   Soon to me was clearly known.
And to me a key was given
By Uriel—guard of Heaven—
Which made every tongue and language
   Of the universe my own.

Then the lights began to glimmer
And their rays grew dimmer, dimmer,
As we sought the Panacea
   Thro' a weirdly dismal damp,
Where all light in gloom was smothered,
Save the ghost-lights there uncovered
From the weird cave of Aladdin
   Where burned his magic lamp.

Said Ariel, "By his magic
We can pass the dragon, Tragic,
And behold the Seven Sleepers
   Guarded by great Al Rakim.
There the cavern will be lighted
And the spirit souls, benighted,
Will dispel the sombre spectres
   By a thought transforming them."

Then we met a hideous sibyl
Who with many a quirk and quibble
Snuffed our light and hoarsely muttered:
  "You who seem so wondrous wise
Go now thro' the darkness streaming
Where there is nor ray, nor gleaming,
Save the glaring of the staring
  Of grim monster's burning eyes."

Here a thousand imps I numbered—
Squabbling imps that never slumbered—
And their eyes were burning, burning,
  Like a glare from lava seas!
Shuddered I, because they knew me
And I feared lest they'd pursue me—
Oh! those slimy sleepless vigils
  Of the world's woes and disease!

From their number, Dissipation,
He who demonized each nation
And who crushed and conquered Reason,
  Rose and proffered me a bowl.
Ariel cried: Forbear to take it
Dash it from you! break it! break it!
For as Undine's kiss 'tis deadly
  Damnatory to your soul!

On we wandered to a roomy
Cavern, where amid the gloomy
Shadows, Yoga, in seclusion
   Took on super-human power.
He divined all earthly dealings;
Read our secret thoughts and feelings,
And he said the Panacea
   Came not of an earthly flower.

Then a fear—a strange fear—seized me,
But Ariel soon appeased me—
On her bosom gleamed the sakhrat
   The most sacred stone e'er found,
And I knew that all the demons
And phantasmagoric lemans
Could not lose us in the cavern
   Tho' a thousand ways it wound!

Soon we reached the Twenty Weepers
Watching o'er the Seven Sleepers.
Round their throats their hair disheveled
   Clustered in a matted heap.
From their eyes almost ideal—
From mortalis lachrymeal,
Flowed their tears unto the dismal
   Scoriac regions of the deep.

All along the shores there staggered
Many souls, gaunt scarred and haggard,
Seeking for the rest to be theirs
   Just beyond the twilight door:
When the spirit Nox came o'er them,
Seized their fainting souls and bore them
For the night to sweet Oblivion's
   Far and fairy haunted shore.

"Come," Ariel said; "We'll follow
To Oblivion's perfumed hollow,
Where the silence broods eternal
   Like the blessed angel, Truth."
And 'twas there, in its completeness
Of perfection and rare sweetness,
That I found the Panacea
   And the sacred spring of youth.

THE MUSE.

I saw her with the cloud queens that hugged and kissed the moon,
  While she grasped a thread of star-shine and followed it through space
  To where water lillies sparkled like bubbles in a vase,
And sighed their sweetest odors o'er the mossy-fringed lagoon,
Where swans in dreamy languor floated to some lulling tune,
  And the leering leaden spiders gave the water-flies a chase.

And I saw her when the dawning winked at day, and she awoke
  And bathed her face in snow-drops, and powdered it with light,
  And feasted on the fragrance of honeysuckles white,
And strolled around the orchard, in her morning-glory cloak,
With an apron neatly woven out of skeins of mist and smoke,
  Appareled like a fairy from the love-isles of Delight.

And I saw her comb her tresses with a thistle-bur that grew
   In the secret isles of sorcery, of Elfland's dreamy dell;
And the mirror that reflected her enchantments like a spell
Was a tiny timid tulip that hugged a drop of dew;
And as she there admired her grace, the purple hyacinths blew
   Their dainty fragrance in her face until they swooned and fell.

And I saw her moulding zephyrs from the pansy's sweetest sighs,
   And sow them, while the fire-flies made melody of gloom,
   And the universe seemed floating in an ether of perfume;
And the moments sped a quiver with the crickets' doleful cries,
While fancy trimmed her vesture with the wings of butter-flies,
   And martial bees held revel in the calyx of a bloom.

And I saw her strip the colors from the sea-shells, hidden low,
   In the dark and dismal cellars of the sighing, sobbing sea;
   And with her fairy fingers skilled to mystic masonry,
She builded arches, rarer than the heaven's arched bow;
While naiads say their sweetest under coral mistletoe,
And the muses swam still deeper in the depths of mystery.

And she stroked the beard of winter with the soft
    of spring,
And he drooped his head in silence and smiled himself to
    sleep;
While clowns of mist, and sunshine came stealing down
    the steep,
And set the flowers to laughing; and birds upon the wing
Ever singing, wove their hammocks with the oriole to swing,
And watch the gray-squirrels gambol, as from limb to limb
    they'd leap.

A GRIEF AND A BLESSING.

In the trail of the star-light I gazed up to heaven,
    One night as I mused on the bank of a stream,
When lone thro' the darkness I saw something driven,
    That fluttered and quivered and whitened the gleam
Of the star-ray it followed, till o'er me it hovered
    And fanned my pale cheeks with its snow-lighted wings,
And it moaned in my ear till my vision discovered
    The cause of its grief and the source of its stings.

In its breast stuck an arrow, a deep poisoned arrow,
    In its beak was a lily from gardens divine,
And it dropped me its lily and sang me its sorrow,
    An anguish I felt that must ever be mine,
And I plucked from its bosom the dart most distressing,
    Deep tarnished with sin as it soared up above,
The arrow a grief and the lily a blessing
    Were all that it left me, God's white sacred dove.

SPIRIT OF DEVOTION.

Come to me while dews are falling,
And the night birds sweetly calling,
　And the whip-poor-will sings sadly
　　In the distant brake and brae;
While the twilight, soft and tender,
Hangs its tinted robes in splendor
　Over nature, hushed and holy,
　　At the dying hour of day.

Come while in the mazy arches
Stars take up their silent marches,
　And the mellow misty moon-beams
　　Spread a glimmer o'er the lea,
And the katydids and crickets
Chirp their vespers in the thickets,
　And the nymphs with pearly dew-pails
　　Scatter mists o'er land and sea.

Come, while queenly night, confessing,
Gains from priestly day the blessing
  That reanimates her chilled heart,
    Throbbing, wrapped in tear-moist dress,
And from beds of musk come creeping
Zephyr fairies, smiling, sweeping
  Perfumes from the dozing roses
    With their light wings' soft caress.

Stainless spirit of contrition,
Listen to my heart's petition,
  As I raise my eyes before thee
    Moist with penitential grief,
And implore the angel writing
In that great book, while inditing,
  To erase the sins recorded
    There against me on each leaf.

A PLEA FOR FRIENDSHIP.

Unknown friends so far away,
If I send to you this lay,
Written at the muses' play,
Will you blush and think me wrong
For intruding thus my song
Where Green mountain poets throng?

Or will you its friendship prize
Like the dainty blue that lies
Dreaming in a violet's eyes?
When you read it, will you be
Like the pippin on the tree,
Blushing back the sunlight free?

If so, friendship will be ours
Wreathed and crowned with posey's flowers
While we dwell in fancy's bowers!
Yes, and more, if we should meet;
Friends until the frost and sleet
Of old age checks life's retreat!

THE SEPARATION.

I know that when she went from me hope's flowers all faded out,
And where they'd bloomed there only grew the tangled weeds of doubt,
And thorns and thistles frowned on me where late the roses smiled.
Nor did I hear the bird's sweet songs again from out the wild
As when we listened to their lays in days forever past,
And talked of love and asked our hearts if those sweet joys might last.

The star-rays seemed to tangle in the night mists o'er my head,
And the moon-beams moaned and beckoned ghostly shadows from the dead,
And the sunlight fell as chilly as a cold December day.
Nothing wore the same bright aspect after she had gone away,
For she had entranced all nature by her beauty, grace and song,
And in going she had taken all their charms with her along.

Never was a night so dreary but some star held out its light;
Never was a wrong so cruel that it did not bring some right,
Never a wind so wildly raging but it blew somebody good,
And in spite of all our grieving things will end just as they
    should.
So it was when I quit drinking and Louise came back to me,
Bringing with her flowers and sunshine that I love so much
    to see.

SONGS OF CHILDHOOD.

Softly on the night winds creeping,
As I lay in dreamless sleeping
Came a tinkling, tinkling, cheering
   Nymph of melody divine;
And her honeyed lips kept nipping,
At my ear and dripping, dripping
Note on note, then raised my eyelids
   And her eyes smiled into mine.

How enchanted then I listened,
As the falling starlight glistened,
And the moon's enamored spirit
   Stared and stroked my window-pane;
And the honeysuckles tilted
O'er their cups as music lilted,
And the roses clapped their lives out,
   At each subtle, stirring strain.

Every sympathy appealing,
To my heart brought some sweet feeling,
And I dozed off into Dreamland,
   While my soul soared with her lays;
And I dreamed of flowery hedges,
Purling brooks and mossy ledges,
And I blessed the nymph for bringing
   Back the songs of childhood days.

MEDITATION.

I am sitting here to-night,
      While the white
      Fire light
Curls and dances like a sprite
      From the fire;
And I watch each ember part,
As the flame-sword strikes its heart,
And its spirit like a dart,
      Speeds from sight.

And the ghosts of other days
      Seem to gaze
      From the blaze,
And to whisper softest lays
      Of the past;
Till the flame has left the coal,
And I see no more its soul,
As the ashes downward roll,
      In the haze.

And I sigh, for well I know,
      On Time's flow
      Love will go
Like fires that smoulder low,
      In my grate;
And to ashes friends will turn,
And their hearts no longer burn,
In that cold and silent urn,
      'Neath the snow.

DECEPTION.

We met in the darkness, she gave me a light—
  A light that gleams only from soft dreamy eyes,
Down into the heart where it burns out the night
  Shadows, that hover around it and rise
O'er the ghost of dead hope and the scenes of the past,
  And then we were parted, she took it and fled,
And I gazed at the spear-pointed flame, till at last
  It passed like a moan through the lips of the dead.

She gave me of honey and nectar, which drips
  Distilled from her heart's fairest rose-words and smiles
And the perfumes around her were breathed from the lips
  Of flowers that gladdened her soul's dreamy isles.
But at last in my goblet the wormwood she wrung.
  Her smiles turned to scorn and her words into hate,
And her eyes thrust a dagger-like look, and it stung
  The angel that guarded my heart's golden gate.

CALL THEM BACK.

Call them back, those scenes of childhood,
Flowery lanes, and tangled wild wood,
Rusting chains of friendship broken,
Cruel looks and harsh words spoken;
Call them back, and teach your heart
What it is to ache and smart.

Call them back, those rays that lingered
O'er the harp your fancy fingered,
Till your dead hopes rose from sleeping,
At each arioso sweeping,
Like some sweet, angelic song;
Call them, bid them linger long.

Call them back, each smile and blessing,
Each warm touch of love carressing,
Each bright meteor spent while falling,
Each lone bird that's sweetly calling
For its mate from out the dell;
Call them back with you to dwell.

Call them back on memory's pinions,
From the silent, dark dominions
Of the past, and ever cherish
Good, and let the evil perish;
Call and let your heart be glad,
Love the pure and scorn the bad.

TRUTH AND ERROR.

As war in Heaven fiercer grew
   And angels fought with swords of flame,
The angel Error downward flew,
   Toward the earth, to scatter shame,
O'er land and sea, and stint the mind,
And dwarf the soul of human kind.
But Truth beheld him in his flight,
And sped behind him with her light,
And as she neared, he backward flung
His spear, barbed like a viper's tongue,
When she applied her torch and burned
Its point until it drooped and turned.
He shrieked, and whet his bat-like wings,
With such tremendous force of might,
Her torch was snuffed out into night.

Then winging back toward the gate,
   Through which she'd sought her way to earth,
She met Oppression, Error's mate,
   Who cursed and ridiculed her worth;

And just behind him Justice came,
Bearing her scale of gold, aflame,
And grasping in her hand a chain,
  With which to bind the tyrant low;
But on her foe she could not gain!
  Her flight, alas! was far to slow,
Yet down to earth she came, and when
She sought the inner souls of men,
She learned Oppression there had been
And left the writings of his sin.

Truth at the gate looked back, and lo,
  She saw each evil spirit hurled
Out from the sight of God, to go
  Moaning through night to another world;
To an outer world, a world of man,
Where each might lay some subtle plan,
To inflame the heart and ensnare the feet,
Of those along life's flowered retreat.
And then she passed the gate ajar
And gazed o'er Heaven near and far,
But not a spirit there remained
Who'd fought against the crown unstained,
And chanting anthems of release
The gates of Heaven closed in peace.

EDGAR WELLTON COOLEY.

Earth fashioned his being, but God lit his soul
    With the supernal rays of a spirit divine,
And gave him to drink from Minerva's pink bowl,
    And builded his intellect like to a mine
Of diamonds and rubies and sapphires and gold,
    That would be revealed to the wonder of men,
On the tablets of time; these he doth unfold
    By the magical genius and gift of his pen.

Each pulse of his heart bears a ray of the sun's,
    Each thought is the kindling beam of a star;
His soul sweeps the universe over and runs
    Thro' each planet that flames in the heavens afar.
His mind probes the most secret depths of the deep,
    And brings up its jewels revealed to the light;
He awakens the naiads and mermaids asleep,
    To sing the smooth rhythm his pen doth indite.

His meter is marked by the mocking-bird's note
    And the musical murmuring stream as it flows,
And its sweetness is likened to perfumes that float
    From the garments that mantle the hyacinth and rose;
So sweet that all nature enchanted looks up
    To the God-given source attuning all spheres,
And the souls thirsting lips plead again for a sup
    Of melody's dew-drops that fall on our ears.

He mounts his Pegasus and rides to the moon,
    And looks back at earth and the stars up above,
And smiles at their comets, and spirits that croon
    O'er the forecasts of life, and the languor of love.
He dips from the cloud-pools the lightnings red ink,
    And stoops down and writes on the mountain's white crest,
Then spurs into darkness and cuts every link
    That shackles the light, with his pen, and is blest.

IS THERE A SWEETER STRAIN?

It is said, dear friend, there's a sweeter strain
   Than the notes of friendship's song.
But I've waited and listened for that refrain
'Till the brooding silence has brought me pain,
   And I think the saying wrong.
For what is love but a strange sweet Flower
   That withers and droops and dies
By the kiss of the sun and April shower?
A thing that thrills the heart for an hour,
   And then its glory dies.

BURNING OF THE STEAMER WELLINGHAM.

Slowly sought the sun, his pillow, and the twilight, o'er each billow
   Hovered like a drooping willow hovers o'er the dead's decay;
Softly, through the ether streaming, came the starlight's mystic beaming,
   And the moon, in guileless seeming, hung her lamp above the bay.

As the boat steamed up the river, wakening each rippling quiver
   To a weird, fantastic shiver, shuddering from the steamer's keel;
While the engine's tireless sighing, floated gently off replying
   To the waves, as they fled flying from the splashing of the wheel;

From the cabin, brightly glowing, music's sweetest strains came flowing.
  Then the smiles, on faces showing hope and happiness and glee,
And the eyes grow dull to pleasure, closed in slumber's dreamy measure—
  Closed to dream of some fair treasure, that in life they'd never see.

In his chair, the clerk sat dozing, in a broken sleep reposing,
  And .the clock ticked off the closing of the midnight's gloomy hour,
When suddenly, the pilot weary, at his post in silence dreary,
  Blew the danger signal, clearly, with the force of frenzy's power.

As a beast intent on killing, as a fiend for carnage willing,
  Sped the red-tongued monster, filling every crevice with his fire!
Rushing madly, clawing, gnashing, twisting, lifting, curling, lashing,
  With his gilded flame-thong flashing, like the lightning's seething ire!

Life to him was but a trifle, as he leaped with joy to rifle
 That fair bark, that night, and stifle many a panic-stricken soul!
And I hear me yet, their groaning, and their piteous prayers and moaning,
 Like a weird, unearthly droning, from some deep Plutonian goal!

Oh! what mind, with wisdom teeming, can depict the awful seeming
 Of those hopeless mortals screaming, as the flames around them swept!
And they plunged into the river, with their frightened eyes aquiver
 With a wild glare, when death's shiver chilled—and they forever slept!

THE QUARTER THAT HE SAVES.

Man may fashion to his fancy each stone in the wall of life;
May fight the many battles that are known to earthly strife;
May sail the seas of fiction, and roam poesy's dreamy isles;
May scale the mystic mountains, and delve deep in history's piles;
Dine with science and invention, and measure mystery's miles,
But the greatest secret to be solved, from the cradle to the grave,
Is the art of making money, and the quarters he can save.

There is nothing in existence higher praised than man's success.
Failure never meets approval, as it wears a tattered dress,
And he who crouches in the darkness with his empty hands at rest,
Fearing to go forth and battle with the world, makes life a jest.
Yet he, who seeks companions that he may choose the best,
May search the wide world over for the comrade that he craves,
But will find no friend that's truer than the quarter that he saves.

IN THE DARK.

Hushed was the hour and the great golden sea of day
   Had paused, while drifted in each warm, each pulsing ray of light.
Darkness descended in her dreary perfumed robes
   And wrote her name upon the ebon scroll of night.
The stars lay drunken in the dreamy, drowsy mist
   And had not lit their beacons; and the moon, it seemed,
Had sent her messengers for light, and they were lost.
   Darkness reigned queen alone, while nature slept and dreamed.

A dim light's flickering glare lit up my lonely room,
   As I sat musing over life's mysterious chain:—
How it was fashioned, burnished, or made dull and cold,
   With here a link of pleasure, there a link of pain.
I heard the voice of footsteps passing by my door,
   That sounded like sweet music to my ear, and then,
No sound was heard, save but the dirge the night winds sing,
   And my soul grew sad and longed for day again.

"It may be that this holds a jewel from the sun,
  Sent by some fairy's hand to light my gloomy mind;"
I thought, as I unsealed a dainty envelope,
  And gazed upon a neatly folded sheet, to find
Three words alone, which wrapped another mystery
  Around night's flapping sails, and doubt's flood-worn ark,
And then I said, "Oh, sorcerers and seers of old,
  Solve me the meaning of these three words—"In the dark."

But in her black pavilion, sulky silence sat
  And opened not her lips; not even was a thought
Sent out to kindle mind with her supernal flames.
  And yet, to solve the meaning of these words, I sought
In vain, until, o'er maddened, I blew out the light,
  And then the mystery unfolded like a pink
Unrolls its blushing petals to the kiss of day—
  For words unseen now glowed in phosphorescent ink.

MAN.

Oh, man, how weak thou art; how frail in all,
  Save to aspire and live and count thy days.
Thy castles are mere fabrics, soon to fall
  And sink amid approval, blame or praise.
Hope lifts thee up; ambition hears thy call,
  And yet thy ways are not God's wondrous ways,
Because thou canst not feel nor understand
The mighty works of His mysterious hand.

A child of progress, yet a thing of shame;
  Sublime, ignoble; lofty and yet low;
A spark between two worlds; a transient flame
  That glows intense, but flits and flickers so,
By it one scarce has time to pen his name;
  A cloud o'er which is but the sacréd bow,
That gilds the storm, and as we view the light,
We sigh, alas! to note its speedy flight.

Truthful, yet false; a coward, and yet brave;
   True, and betraying still; a friend, a foe;
A worm, a god; philosopher, a knave;
   Known, yet hidden; ripe, yet in the dough;
A wonder-working prince to feed the grave;
   A shadow, moving thither to and fro,
That leaves a trace of something lit of mind,
Then vanishes in darkness, dim and blind.

Thou art an atom thirsty with desire,
   That never dies, though time may thwart thy aims,
And quench at last ambition's fitful fire,
   That stirs the soul of youth to deeper flames.
With thee came gain and loss; pleasure and ire;
   Passion and pride and wealth; yet nature tames
Thee on her bosom. But an infant care
Thou art, as when first taught thy childish prayer.

And thou shalt love and worship, and yet lose;
   Soar, and sink back; arise and sink again;
Fashion to fancy, seek and find; and choose,
   Yet all will vanish, leaving thee but pain.
About thee, fate entwines her fatal noose,
   And moves thee on with nature's noiseless train.
Thy tears, and hopes, and prayers can nought avail
Death's ear is deaf to man's poor wretched wail.

Thy heart may yearn to move the mighty world;
　　To calm the winds, and still the rolling tide.
Yet, in the silence, they are onward hurled,
　　Leaving thy hopeful heart unsatisfied.
Not by thy skill was bud by dew impearled;
　　Not by thy might does blazing Pheolus ride,
But thou art as the insect on the crust
Of mother earth, thou treadest into dust.

Thou hast a place, yet not a throne on earth,
　　Nor on the orbs that circle round the sun,
For they in beauty, brilliance, splendor, worth,
　　Rolled many cycles ere thy race begun!
The vast creation knoweth not thy birth.
　　Though empires fall, and cities sleep undone;
Thou hast no part in nature's stir and strife—
Thy voice spake not the universe to life.

Thou didst not give the seas their jeweled beds,
　　Nor fill the vaulted sky with ether waves;
Nor cap with snow the mountains' lofty heads;
　　Nor goad the dreadful crater till it raves.
But thou canst feel the force that nature sheds.
　　And thou canst strew the earth with myriad graves,
Just as thou wasteth, nature wasteth thee.
Vain, boasting man! How brief thy time to be!

Thou canst not chain the lightnings in their pride;
   Nor hush the dreaded thunder's sullen roar;
Nor stay the imperious avalanche's slide;
   Nor lock the hurricane within thy door.
By all, save beast alone, thou art defied,
   Although thy thoughts to distant worlds may soar;
Still thou art lost, because thou canst not feel
And understand what nature doth reveal.

Fate gives thee birth, and mankind lifts thee up,
   Or lowers thee, as it doth please him best;
Sorrow and joy alike doth fill thy cup,
   From which thou drinkest either cursed or blest.
Full many a flowing nectar thou may'st sup;
   Full many a pang to give thy soul unrest.
All thou canst do, is live, seek, slave and sigh;
And follow after beckoning hopes—and die.

To die? Ah yes, what is it but a sleep,
   Wherein we loose this heavy clog of clay?
A pleasant dream, from which we wake to keep
   A spirit form through all eternal day?
The grave should hold no fear to make men weep,
   As it, to higher life, doth point the way,
Where every pang and plague of earthly strife,
Is swallowed up in sweetest celestial life.

THE VISIBLE PANDEMONIUM.

The Hell I see is not where spirits band
    Together in some torturing place of gloom,
Beyond the grave, where fiends satanic stand
    And plead for respite from a burning tomb,
But here on earth there lies on every hand
    Fell Erebus, that brings to each, unpitying doom,
And makes us wish that right, not wrong, would win
Each human soul from suffering deep, and sin.

Go view the dens where man his revels holds,
    And swaggards quaff damnations liquid flame;
Where wild delirium wraps him in its folds,
    While like a thief he fouls his Maker's name.
Men crazed by drink, when all the story's told,
    Have strewn the land with woe, want, death and shame.
In this a hell I see, deep, ruinous and dire,
Where is no lake of burning brimstone fire.

Where Prostitution flaunts her shameless face,
  Do not the throes of Hell engender there?
There loathsome death and pestilence embrace,
  And future ages these vile horrors share?
The lustful soul raped thus of every grace,
  Is left a putrid mass in dark despair.
Is not this Hell? Then teach me sacred verse,
Where I may find a fouler hell, or worse!

The ghoulish ruffian, stained with blood and crime,
  Who shuns the light, and seeks night's gloomy scroll,
Who, like a snake, crawls through the filth and slime,
  On murder bent! So dwarfed his meagre soul,
That, dragged by force from out his beastly grime,
  He'd scurry back to some obscurer hole!
Tell me, ye priests, I pray you truly tell,
Is truth, and love, and decency his hell?

The bawd procurer, lustful, seeks his prey,
  His chiefest aim, some virtuous life to blast;
Disowns his conscience, throws all good away,
  While villainy and falsehood, overcast
The truer light of manhood's perfect day;
  A fouler whelp of vice can not be classed
With men. A brute so whelmed with purpose, fell,
But seeks to make of earth a noisome hell.

Within the dark and solemn, silent hours,
    While innocence and virtue lie asleep,
The abject miser, meanly, basely cowers,
    And holds communion with his hoarded heap.
The coins, to him, have strange, mysterious powers,
    And sink his small and niggard soul so deep,
That, palsied by their sick'ning, baleful spell,
He, dog-like, revels in his yellow hell.

In prison cells, the wretched convicts wait,
    Hedged in with walls of brick and steel and stone;
Shut from the world by the strong hand of fate,
    Where no kind voice responds to sorrow's moan,
Till weariness and grief insatiate,
    Take all but life, and that left drear and lone;
For those, who thus with silent horrors dwell,
There can exist elsewhere no viler hell.

The judge, in ermine, fringed and stuffed with gold;
    The court officials, smirched with lucre's stains;
The lawyers and the jurors, bought and sold,
    Pit man 'gainst man, for base, ill-gotten gains.
Twixt life and death, the innocent they hold,
    Within some cell bound down by legal chains,
Until e'en thought grows sick, that knows so well,
That these foul devils' kennels are but hell!

Where nodding trees swing restless to and fro,
   Across the path the whispering winds have made;
Lone, lorn, and desolate, and marked with woe,
   The pest-house stands beneath their silent shade.
Above its walls the reeking odors blow,
   And noxious death befouls the slumb'ring glade,
Proclaiming loud to God and man as well,
That earth holds many a weirdly, loathsome hell.

In these sin-crimsoned scenes on every hand,
   I trace the shadows low'ring o'er life's road;
I hear and see; I feel and understand,
   That villainy still seeks each vile abode,
And while within emotions, true and grand,
   Live yet, the heart would gladly void its load—
Existence fain would break its filmy shell,
   And scape by death, each cursed gloomy hell.

WHERE THE BATTLE WAS FOUGHT.

When the earth at dawn awakened in the hammock of her dreams,
    And the sun turned back the coverlet of darkness from her brow,
She shivered as the tempest hid the sun's enamored beams
    And wrapped her in its mantle, and kissed her with a vow.

A soft, white, fleecy vesture, unruffled and unstained,
    As fair as any lily that e'er kissed the lips of light,
Was her snow-robe on that morning the war-worn soldiers gained
    Their position for the conflict and gazed along the height.

When the dusky queen of dreamland scattered darkness from her wings,
    And the moon hung out in heaven her dreamy, lurid light,
She sighed to see earth's garment deeply stained and torn in strings,
    And swooned in an eclipse from the soul-appalling sight.

THE HEAD LINES OF A LETTER.

Grim night her sable curtain spreads across the world,
   While sighing winds their doleful tales repeat;
The dews descend, and every flower's impearled
   With nature's breath, the life of all that's sweet;
The gloaming's past; the vesper's song has died,
   The moon's soft dimpled face steals o'er the hills,
While in their cars, the blazing planets ride,
   And slowly grind the gods' unceasing mills.

And I sit here within my lonely room,
   Tracing the thread that winds existence down;
No friend to cheer, no smile to light the gloom,
   While fate and fortune deign on me to frown—
Frown, while hope's angel, smiling, strokes the face
   Of darkness with her snowy, sunlit wings;
And sprinkles wine from out a jeweled vase,
   To heal the heart that melancholy stings.

My thoughts alone are all that I possess,
    And, were they set aflame in poesy's urn,
Me-thinks they would be quenched in wretchedness,
    Before one living eye would see them burn;
Oh, why this deathless thirst for fame and power,
    When all the world laughs at our bootless aim,
And friendless, homeless, moneyless we cower
    Beneath its jest, and crawl to win a name?

A LESSON IN MYTHOLOGY.

In the hall of the gods one still summer night,
   Where roses and lillies breathed sweetest perfumes,
A goddess appeared, in her mantle of light,
   Wearing buds sweeter still than the god's choicest blooms.
And when they beheld her enrapturing face,
   That glowed like dew-pearls in a lily's white throat,
Apollo, enamored by her magic grace,
   Struck his lyre's gold chords, while note after note
Dripped off the strings in a rhythmical stream,
Like the perfumes that drip from a rose in its dream.

This charmed the sweet goddess, whose cheeks were as fair
   As snow-misted marble revealed to the light,
And stroking the threads of her mystical hair,
   She danced to the notes like a mythical sprite.
Then smiling, she gave young Apollo a pink,
   That bloomed in the warm pulsing heart of the moon,
Around which was woven a chain, link by link,
   From the dainty gold beads she had plucked from his tune.

Then dissensions arose and the gods angry grew,
    To think that Apollo had won such a prize,
Till Hercules wrathfully seized her and threw
    Her out in the fathomless mists of the skies.
Then Jupiter put in his jewels and swore,
    And called upon Phoebus to check her dread flight,
When her hair, and the long trailing raiments she wore
    Were instantly turned to a mantle of light.

But they could not recall the rash act that was done,
    And as jealousy raged so intense in the hall,
Each god took an oath with the god of the sun,
    That she have no rest, but eternal must fall
Through space, round and round each planet that rolled,
    Till the universe melted and sank into night;
And they called her just Comet, because of the gold
    Tinted robes that she wore and the speed of her flight.

CUSTER'S LAST BATTLE.

Tired and worn the forces halted,
    While the low, descending sun
Sank behind the towering mountains,
    As the busy day was done.
And the camp-fires lit the valley
    With a glimmering, ghostly light,
As they faded in the darkness
    Of the hushed and solemn night.

Then the midnight's deathly shadows
    Sealed their drowsy lids in sleep;
And the breeze from off the mountains
    Seemed to softly sob and weep,
Through the branches of the forest,
    As the darkness wore away,
And the silent stars and sentries
    Held their vigils until day.

As the dawning day unfolded
   Light above the mountain peaks,
Horror stupified their visions;
   Terror blanched their glowing cheeks,
As the painted savage legions
   Hedged the mountain sides and dells,
And the cliffs and forests trembled
   At their dread and hideous yells.

As a Jupiter, when maddened,
   With the lightnings at his back,
Hurls the thunder-bolts of heaven
   Fierce and wild along his track,
So they rushed in furious hatred
   To the wild tumultuous fray,
Like a hissing prairie fire,
   That spares nothing on its way.

When the sun had climbed the mountains,
   Death-like shades obscured the light,
For the sullen war-cloud bound it
   In its blood-stained folds of night,
And this sad and solemn message
   Floated over hill and plain;
"In the valley of the Big Horn,
   Custer and his men lie slain."

Still the snow-caps dress the mountains;
   Still the valley blooms with flowers,
And the little Big Horn river
   Flows eternal with the hours;
Still the star-beams shine forever,
   Spectral sentinels keeping guard
O'er the slumbers of the soldiers,
   Sleeping underneath the sward!

WHEN IT SNOWS.

The night retires; young day awakes in chill;
Snowflakes descend o'er city, vale and hill,
Lightly rifting, softly sifting through the dawning o'er
    the land,
As if they were rose leaves drifting from an unseen
    angel's hand.
Blue, through the mist, the smoky currents rise,
Commingling with the white flakes from the skies.

The brown earth powders white her dusky face,
And smiles in dreamy fancy at her grace,
Brighter growing, as the flowing flakes fall silently and
    light,
In the north wind drifting, blowing, till they hide her
    dimples white.
Then like a bride in virgin beauty sweet,
She slumbers folded in her milk white sheet.

Along the stream the solemn forest stands,
Catching the chilly fleece with naked hands,
Bowing, dreaming, at the seeming loveliness of its own
    dress,
Till the sun's impassioned beaming burns the fleece to
    nothingness.
Then it lifts its hoary head with many sighs,
While snow-tears melt and trickle from its eyes.

THE DREAMERS.

This world's a rusty riddle; an interwoven net,
That entangles and confines the mind in mystery, and yet,
Through every mazy labyrinth, we seek and search to find
Some jeweled theory, to throw new light into the mind.
And every time the earth revolves, some sage comes bounding out,
With his dark lantern, shouting, "Eureka, do not doubt."
We canter to the customs of every turn and phase,
And weakly, blindly follow each advocated craze,
No matter how denying to the human race it seems—
The world is filled with dreaming, and the dreamer tells his dreams!

Oh, the myths, and creeds, and dogmas, thrown from the orb of night,
Glint through the mist of reason, but in Superstition's light,
Quite confound the dreamy vision, while each stale hypothesis,
Like a scepter, comes before us from some fathomless abyss,
Till Philosophy, Doubt's armorbearer, like a slave set free,
Breaks through the darkness with his torch, that we may clearly see
The face of Truth, as she is seen through Nature's wondrous plan,
So long obscured in darkness, by the myths and creeds of man.

THE WAR.

It has gone with its terrors, and anguish, and death,
Like the cyclone's mad sweep, or the Simoon's hot breath;
It has gone, and peace reigns over mountain and plain,
Where the green tufted mounds make a note of the slain;
It has gone, but its ghost stalks on sea and on shore,
In gore-tinted robes, midst the gloom evermore.

How well we recall when the red eye of war
Flashed vengeance and death to our kin near and far;
And camp-fires shone over each battle worn host,
While the vigilant sentinel stood at his post,
Till the soldier awoke from some pitiful dream,
But to bend o'er his comrade and watch his life stream
Ebb away, as the tide pouring over a steep,
Rushes on and is lost in the fathomless deep.

No longer our cities are fed to the fiend
Of war—for its terrible harvest is gleaned;
No more civic slaughter, the earth's foulest stain,
That man ever witnessed on land or on main,
But the goddess of freedom wears peace as her crown,
And her light hides the anger of War's sullen frown.

IN A WORLD BORN TO DOUBT.

In a world born to doubt, I went seeking one day
A secret unknown to the giddy and gay,
And as down the streets of the city I went,
Meeting faces that seemingly glowed with content;
Meeting faces as fair as enchantment e'er told,
And others that stared, all expressionless, cold.
I questioned each look, each glance of the eye,
But the secret lay hid in the heart of a sigh,
And a spirit that answers the soul's earnest call
Whispered soft as a zephyr's: "The face tells not all."

I stood on the bank of the ocean and gazed,
As day on the altar of night dimly blazed,
And I sighed, as the last fringy flickering light
Faded into the deep, melted into the night;
While star-gleams in love kissed the hills and the sea,
And the waves seemed to echo a sweet lullaby,
And the moon in her bark o'er the ether waves smiled,
As a mother would smile o'er her slumbering child,
While a voice, falling soft as the night-shades do fall,
Said: "Know ye, O Mortal, Love tells not all?"

"There are secrets that sleep in the womb of the earth,
Which in no mortal mind will ever have birth.
Each face that you meet is nor compass nor chart
Of the deep lying seas of the soul's inner heart,
Each star holds a secret as dark as a tomb,
And the sun's burning light to the mind is but gloom.
The air floating round you, the waters that roll,
The flowers that bloom, and the thoughts of your soul
Are murmurs which echo through mystery's hall,
Moaning ever and ever, 'Life tells not all.'"

And then while the night rocked the ocean, a storm
Rose out of its bosom which heaved in alarm,
And I saw on the weird, maddened waves, rolling high,
A vessel that tossed in the lightning's red eye,
Like a soul that is lost in a city of souls,
Till ravished it lay on the pitiless shoals,
And as the mad sea wrung its hands in affright,
There came a loud voice falling out of the night—
A voice falling dark as a funeral pall,
Crying: "Death, foolish man, alone telleth all."

'TIS SWEET.

'Tis sweet when perfumed zephyrs mildly creep
    From leaf to leaf, from budding pink to rose,
To sit and gaze far through the ether deep,
    And watch the moon and stars wake from repose.
'Tis sweet by music to be lulled to sleep,
    And as your drowsy eyelids droop and close,
Float in your dreams to some elysian isle,
Where every sound is love, and every bud a smile.

'Tis sweet to sit upon the lonely shore,
    Free from all care, and hear the murmuring deep,
And watch the sea-gulls flitting on and o'er,
    Like spectres of the air that never sleep.
'Tis sweet to scent the salty breeze and pour
    Your dreamy meditations on each heap
Of ocean's treasures, and the gems that lie
Revealed alone unto the mermaid's eye.

'Tis sweet to watch the ships come in from sea,
    And know some loved one soon shall have returned;
Sweet is the thought of mother, and to be
    With her whose heart for you so long hath yearned;
'Tis sweet once more to sit upon her knee,
    As when a child, that infant prayer you learned,
And feel and know that loving spirit still
Abides, unchanged, through every good or ill.

But sweeter 'tis at eve, the sun sunk low,
    When brawny labor seeks his home to rest,
To see, from out his cottage window glow
    A light, to thrill and cheer his honest breast;
'Tis sweet to meet his children and to know,
    With health and comfort, he and his are blest;
Sweet are their smiles, their childish songs and prayers,
Although they've brought to him a thousand cares.

Oh! home of labor, sweet thy walks and ways,
    Where smiling lips await to kiss and greet,
And sinless, baby eyes look up in praise,
    To cheer the gladdened soul of those they meet;
Oh! home of toil, thy hearth, thy firelight's blaze,
    Makes the dark shadows of despair retreat,
Where love and sympathy and joy unite
In one supernal halo of delight.

A SONG OF A JOY RETURNED.

My life was once like a starless night
In its mantled gloom with no ray of light,
So strewn with wrecks of my hopeful years,
So gorged with failure, o'erflowed with tears,
Pursued by storm's relentless sway,
That I scarce had hope of a brighter day.

Who has persuaded, with good effect,
The heartless world to give respect
When censure's eye and venomed tongue
Have poisoned words to world out flung?
Who has not cherished a hope in vain,
To feel but torture; to know but pain?

Oh, sighing sadness, go hide thy face,
Till its furrowed lines I no longer trace.
For joy's bright rays have again returned
Into my gloomy heart and burned
The last dark shadow that hovered there,
Like the pall o'er the altar of despair.

TAKE LIFE AS IT IS.

Take life as it is and dread not the future,
   Though the past you have seen through a veil-cloud of tears.
There's somewhere on earth a loved spot for each creature
   To harvest a sheaf of sweet comfort in years.
Why fret life away like the storm cloud that rises,
   And wastes itself mad sweeping over the plain?
Why not spend it calmly, as voiceless clouds gather,
   And let down the mists of the morning in rain?

Take life as it is; or as God has intended,
   Let the soul soar away from the shadows of night.
Treat the earth as a garment that stain has offended,
   And borrow from Heaven your treasure of light.
The smiles and the tears, the frowns and the sighing,
   Are things that we reap from the broad field of life.
The pain and the pleasure our days are supplying,
   But add to our measure of bliss or of strife.

Yet what do these profit the web that is woven?
    Even friendship's fond vows and the sweet tales of love
Are hollow and empty, unless they are proven
    In fond deeds inspired by the Father above.
Our trials and sorrows may come as a blessing
    Clothed in disguise, as our rank passion schools,
Yet we should strive every day for the better,
    Believing in all that a Providence rules.

You, on the tread-mills of life, worn and weary;
    You, in the dark-hidden mines of the earth;
You, in the chambers of prison so dreary,
    Know peace and plenty may smile at your hearth.
No one can see what the future holds sacred,
    Nor can they measure the strength of their years.
Flowers often grow where the sands have been watered
    By blood of the dying and sorrowing tears.

Soldier, your lot may be hard and ungrateful;
    Lover, your love unpropitous may be;
Sailor, your life may be cast with the hateful
    Storms tossing you on a mad, surging sea.
Yet to the warrior the angel of mercy
    Bearing the olive of peace soon must come,
And with his bride will the once saddened lover
    Watch the brave sailor land safely at home.

Friend to the bowl whose riches are squandered,
    Curse you the idol now sealed to your breast?
Why to the shores of despair have you wandered,
    Since God wisely willed everything to be best?
Life is a union of much dissipation,
    And though the snares have inveigled your feet,
Providence wisely made it His mission
    And left ample room for your saving retreat.

Let hope in the future e'er silence our sadness,
    Though our past has been draped with the veil of regret;
'Tis sinful to spend life in folly and madness,
    Because fortune's light has not dawned on us yet.
For, out of the shadows of night while we slumber,
    The angel of mercy, with all her rich store,
May pause for a moment to smile on us sleeping,
    And leave us a blessing stacked up at our door.

SLANDER.

When will the golden dawn arrive,
    Proclaiming peace on earth,
And Slander and his consorts
    Gasp and strangle at their birth!

Oh God! when will our people
    Be freed from Slander's chain?
And when will Thy true Kingdom
    Upon this dark earth reign?

When will the lips of liars
    Be ever sealed and dumb,
Before the scorn of mortal,
    And the wrath of God to come?

UPWARD.

Spirit, that cleaves the highest height,
With whiteness whiter than the white
Far up beyond the light of light,
      Carry my rod,
And aid my feet that I may climb
      And leave my hope with God.

JEALOUSY.

Oh curse thee, hag, with thy dark spider eyes,
    Too well do I remember neath the moon
The time I met thee and esteemed thee wise
    And hearkened to thy witchery and croon,
Not dreaming that thou sported Psyche's guise
    As splendid as the virgin sighs of June.
Not dreaming that in thy perfumed robes there toyed
And coiled the adder of all peace destroyed.

Not thinking, that when first with thee I sipped
    That strange mysterious wine from Vesta's bowers,
That I had quaffed a poison: One that dripped
    In four-fold triturations from the flowers
Of ceaseless agony. I might have dipped
    A pearl-shell ladle into Reason's showers,
And drank, as Bacchus drank himself to sleep,
Nor pang of thine to burn and make me weep.

Accursed the time I drifted in thy bark
    With thee, while Peace knelt down upon the beach,
And sobbed with wild gulls, moaning to the dark,
    As she stretched out her arms in dismal reach,
As reach lost souls to grasp God's saving ark
    From torments deeper than the lips might teach.
O, thou hast made my heart so swart with sin
Love takes affright whene'er she gazes in.

IN SLUMBER LAND.

Night came creeping o'er my dreaming,
   But the dream was sweet to me,
As I wandered where the gleaming
   Star-dust fell into the sea,
And the moon bent down and whispered
   To the waves a fond good-night,
As she bathed her feet and ankles
   In a gleaming bowl of white,
From which dripped in quivering beauty
   Flossy streams of liquid light.

And as Luna sailed off smiling,
   In her golden bark of June;
Fairy forms pursued her, piling
   Wreaths of sombre and maroon
All around her, till the darkness
   Covered up her dreamy hair,
And the mermen chasing after—
   Chasing as but mermen dare—
Gave the water with their torches
   A pure purple, sea-shell glare.

Then in Oriental splendor
   O'er the waves a vessel came,
Decked with fish-scales, and the tender
   Tintings of them cast a flame
Round about it, like the glory
   That enwraps a hedge that's old
In the sun glare of October,
   When in mats of bronze and gold
Autumn stands to greet November,
   Her old frost-crowned lover bold.

Oh, 'twas sweet to me while dreaming
   To behold the vessel's crew,
And their rainbow sails a-streaming,
   Brightened by the falling dew,
And to hear the lilting music,
   And each pulsing undertone,
Floating off from fairy fingers
   And from lily-bugles blown,
While my soul, through heights of fragrance,
   Breathed its way to God's white throne.

THE BANQUET.

In the heart of a populous city
   In a palace of grandeur and art,
Where Luxury's minions are smiling,
   And Fashion holds sway in the heart,
They are holding a banquet to honor
   A man with his millions of gold;
Not thinking of aught save their pleasures,
   As greedy Belshazzer of old.

Where rank and position are welcomed,
   Joy sparkles the wine in its bowl,
While music, fair heavenly maiden,
   Draws love's sweetest notes from her soul.
In vases of china and silver
   Are flowers from a tropical grove,
Whose fragrance, by light winds, is wafted,
   Like the mesmeric odors of love.

Unminding the tears and the sorrow,
　　That saddens the world and the soul,
They sit at their table cajoling
　　To welcome the kiss of the bowl,
As the small, nimble feet of the giddy
　　Speed on with the musical dance,
Hands clasping hands overburdened
　　With diamonds that dazzle the glance.

Near by, in a dark, gloomy garret,
　　A poor widow, wayworn and sore,
Sees her little ones steal from the window
　　To a pallet of straw on the floor.
Their ears had grown dull to the music;
　　Their eager eyes lulled into sleep,
And the mother, with heart overburdened,
　　Bends o'er them, in pity, to weep.

They thought not of sorrow nor sadness,
　　Nor of earth's selfish darkness of night;
As the satellites borrow their sunshine,
　　They witnessed this scene of delight,
Where the damask and purple hung fragrant,
　　With odors of roses and vine,
And gaslights that mimicked the splendor
　　Of a glimpse of a glory divine.

At midnight, patrician and peasant
    Alike heard the clock tell the hour,
One stitching for bread in the garret,
    (The other, in luxury's bower!)
Poor, sickened, and saddened, still sewing
    For a pitiless privilege to stay
And thread out her life on a garment
    To the stroke of the shuttle each day.

At daylight the banquet is over.
    Its mirth and its music are flown,
As sunlight steals into the garret
    To witness a poor, lifeless stone!
In her fingers the shuttle lies empty,
    From the spool she has wasted life's thread,
And the garment is yet to be finished
    To buy her three children their bread.

Oh God! how I pity a people
    Whose aims are so selfish and high!
Oh God, how I pity the wretched,
    Who have to thus struggle and die!
One whirls in the maelstroms of fashion—
    One's lost in adversity's sea;
One lives as a slave unto passion,
    And death sets the other slave free!

One dwells in a heaven of comfort,
    Illumed by a diamond's bright glare—
The other, a servile in darkness,
    The hell of her doom and despair!
Oh, mortal, consider your ending,
    And who in God's sight is of worth—
At last in this world we are equal,
    When mixed with the dust of the earth.

"WHEN LOVELY WOMAN."

When lovely woman stoops and fails to shame,
    Who seeks to know the truth about her fall?
Upon whose heads do Christians heap the blame?
    The curse falls on her own, of course, that's all.
They are the first to execrate her name,
    And mantle it in a disgraceful pall.
From love, and care, and chastity she's hurl'd
By those who keep unspotted from the world.

What of the man who caused her fall and blight?
    How does the "straight-backed class" regard his act?
"The trifling thing, he treated her just right;
    She had but little reason and less tact."
He, perhaps, reforms, becomes a Christian light,
    And prays for scarlet sinners and souls racked.
The church her doors throws wide and takes him in,
But closes them against this "maid of sin."

It does not weigh the villain's subtle lies,
  His powers of passion to subdue the mind,
Nor how, with Judas' kiss, he hourly tries
  Her love and virtue, with his lust to bind.
And if she yields, and thus sweet virtue dies,
  Around her love the serpent is entwined,
Another life to shame but pays the debt,
And all but her live on with no regret.

Another page in life's great book is soiled;
  Another turn at fortune's wheel brings loss;
Another pattern God designed, is spoiled;
  Another virtue nailed on to the cross.
Hope, future, fortune, peace and honor foiled,
  Because man's heart is filled with lustful dross!
The man goes free, the woman lives in shame,
And dies unrecognized, without a name!

A STRANGE FAIR LADY.

What strange fair lady is this I see,
    As she goes to her work each day down town?
I've noticed her glancing oft times at me,
    With never a smile and never a frown.
But her face is as fair as the delicate flower
That blooms unseen in some lonely bower.

Could I read in that face, so faintly fair,
    Deep lines of sorrow, were I to know?
Is her bosom aflame with its earthly care,
    And her life a burden beneath its woe!
Why is it, I read in that sad, sad face,
A picture of sorrow, and still of grace?

For six long years she has passed this way,
    At dewy eve and at early morn,
And never a smile on her lips I see—
    Never a smile as she looks at me,
But her downcast gaze on the pavement, cold,
Leaves a shadowy trace of a tale untold.

A tale untold, like a song unsung,
  Is forever dumb to the curious ear.
It may be aglow with a smile suppressed,
  Or tinged with the glint of an unshed tear.
But, though from a whisper it had its start,
It will live eternal in somebody's heart!

And those who knew her have sighed and guessed
  At the secret within her heart hid low.
But sigh answered sigh, like waves on the breast
  Of the surging sea, at the tide's inflow,
And never, I ween till the last great day,
Will the secret be known, why she's not gay.

A WOMAN'S WORTH.

Amid the vexations and cares and strife of this busy old earth,
How many men do there live who prize a woman's true worth,
As she labors and struggles and frets and worries from year
    unto year,
Only to earn at the last the glow of a moistening tear?

Oh, the deep pangs of her heart and the pains covered up by
    a smile!
Oh, the deep lines of her face that will fade in the grave
    after while!
Oh, the moth hectic that wastes and withers the rose on her
    cheek,
The secrets interred in the shroud of her soul, that she will
    not speak!

The scourge of the unwritten law established to Charity's shame,
That says woman only may live by the right of her virtue and name!
Have those who established this rule of action, for her held the light
And guided her faltering feet through the shadow and gloom of the night?
Or will they look into their hearts' secret closet of rubish and lust,
And find if this virtue of hers is not trampled down into the dust?

Dirt and cobwebs and soot and smoke of a desolate room!
Oh! where is the light of love's torch that is left to burn out the gloom,
Oh! where is the kiss of good-bye and the smile and the kind word to cheer?
And where is the kiss at return that ever to woman is dear?
Drowned but by kiss of the bowl o'er the bar with the vulgar and gay!
A curse is the love she receives for her struggles and cares for the day!

Oh! have you thought how for love she has wasted her life
    all these years?
Oh! have you heard all her prayers and measured her sighs
    and her tears?
As she bent o'er her suffering babes to soothe and to set
    wrongs to rights.
Oh! have you numbered her hours of gloom through the long
    sleepless nights,
Or have you each day, thoughtless man, with your own
    brought a new store
Of shadows, vexations and cares, and frowns to lay at
    her door?

Oh! does she weep or rejoice o'er her own darling girl
    baby's birth—
O'er the life it may live and lead o'er its destiny here on
    earth?
Does she measure its sorrow and joy and failure and strife
    with her own,
And see through the dim, dawning years, the pitfalls dug for
    it when grown?
And out o'er the widening years, as life's day fades into
    a spark,
See the angels of hope speed away to hide evermore in
    the dark?

Oh man, wilt thou cherish and bless and lighten the gloom
    and the trials
Of her who was born to grace and cheer the earth with her
    smiles?
Oh! wilt thou honor the most divine work of the Deity's
    hand,
And give her an altar that's built by virtue, where she may
    stand.

Wilt thou show more mercy, Oh man, and gladden the
    drooping heart
Of her whose life and whose soul forms of thy own a part?
Kind words are so easily said, but they lighten the burdens
    of life—
God knows they are heavy enough to her who is mother
    and wife!

SPRING.

Wandering, like a saddened memory
   From the silent past abysm,
On the silvered shores of winter,
   Floats the golden spring-tide in,
And our hearts pulsate with rapture,
   As we catch the sweet baptism
Of the smiling sun now glowing
   O'er the city and her din.

Oh! the glory of the springtime,
   And its magical completeness,
As we scent the fragrant odors
   From the garden and the ledge,
And we feel our heart-strings quiver
   With the melody of thrushes,
Till our senses swim bewildered,
   With the music of the hedge.

In the flowery-ladened meadows,
   Where the moss-fringed brooklets murmur,
We sit and drink the sweetness
   Of the whip-poor-will's low strains,
And we sigh with joy to listen
   To the tireless, tremulous twitters,
Floating on in rhythmic currents
   Down the maple-scented lanes.

In thy heaven-lit, mystic mirror,
   Lovely spring, before me glowing,
I behold each fragrant blossom
   Of my childhood's sunny bowers—
Blossoms that the queenly summer
   Of my life, has plucked and wasted,
Yet, whose shadowy beauty lingers
   In thy mirror's magic powers.

HOW SWEET TO BE FOLDED AWAY.

When the land marks of life are destroyed,
    And sorrow sits grim at our gate;
When the gold of our hopes is alloyed
    By the dross of deception and hate;
When the chancels of love are defaced,
    And her wines have been cast in the fire,
And hope from our life-page erased,
    What more has our heart to desire?

When the rich blooming gardens of youth
    Have turned into deserts of sand,
And friends have forsaken the truth
    And raised high against us their hand;
When the smiles of affection have changed
    To the frown of malevolent hate,
And we, to all peace, are estranged,
    What more can we question of fate?

When sad, are the faces we meet,
    As we gaze through a curtain of tears,
And the names we would gladly repeat,
    Alas, are unspoken for years,
And we sigh at each parting of day,
    Like a wave that sighs back to the shore.
How sweet to be folded away
    In the bosom of sleep evermore!

PASSION.

CONSTANCY.

O Love, believe so long as I
   Have life and being from life-giving power,
This love I hold for you shall never die,
   But grow and ever deepen, hour on hour,
As time, on tireless wings, flits ever by.
   Not as the fading tint that gilds the flower;
Not as a whisper spent, nor passing sigh,
   Is my true love for you. But O, as deep—
Deep as the deepest depth the world contains;
   And at each heart-throb even tho' I sleep,
Love's fire rekindles life within my veins;
   Love brings thine image for my soul to keep!

And should I pass into death's dismal shade,
   And you should sit and sigh my face to see,
Love, I would tread the darkness unafraid,
   And draw night's starry legions unto me,
That by their lights divine and mystic aid
   You should behold my image true to thee,
And view the soul of love, most perfect made—
   The soul's deep inner soul of Constancy!

A VALENTINE.

Oh, blest is the lover whose love is returned
   In the same divine measure it's given,
As the beams of the sunlight that fall in the lake
   Take back all their lustre to Heaven.
A life without love is a vile, stagnant pool,
   Whose waters but poison the air;
Then Emma, let love from the fount of your soul
   Refreshen this heart of despair.

As the thirst-stricken doe at the river's dry bed
   Imploringly looks to the skies,
So my heart at the stream that affection has fed,
   Is a-thirst for the light of your eyes,
As the sunlight to day—as the stargleam to night;
   As the deep rolling tide to the sea,
Thy love is the substance, the soul of the light
   Hope ever will shed upon me.

HOW SAD IT WAS.

How sad it was from thee to part
  The stricken lark but knows it—
The tear deep hidden in the heart,
  Like the poison arrow shows it.
As the misty clouds of morning lift
  And darkness blends with day,
The soul of my soul's sigh must drift,
  And melt with thine away.

Before thy heart will feel the pain
  That absence from thee brings,
Oh, love must wear the galling chain
  And feel the adder stings;
And day must go, and night must come—
  The night that bringeth grief;
And Love must dwell in darkness, dumb—
  Poor Love, so sore a thief!

Then doubt shall breathe a jealous spark,
    To flame and burn the mind,
Till Hope's lone dove, from out the ark
    Of passion, speeds to find
The olive branch of confidence,
    Plucked from the flood-washed tree,
When from the troubled waters, dense,
    May a rainbow rise for thee.

A LAST GOOD BYE.

Fair Enola, I have loved thee,
   But the fatal fever's o'er,
And the strange, sweet charm that swayed me,
   Ne'er will thrill my dead heart more.
Yet in memory's magic mirror,
   All thy charms before me play,
And thy gentle spirit haunts me,
   As the sunshine haunts the day!

Blindly loved I thee, Enola—
   Loved thee as the lark, the morn!
In return thy heart has cherished
   My love only for thy scorn!
In thy pride didst hate and spurn me,
   As thine eyes and lips oft told,
Thou hast wooed a princely lover—
   Hast betrayed my heart for gold!

Yet some ray of light may glimmer
   O'er the love thou'st cast away,
And my name in memory linger
   On thy lips another day—
Linger, as the sun at twilight
   Lingers, like a parting sigh,
And then, dying, leaves but darkness
   Over him who says, "good bye."

IT IS SAID YET NOT TRUE.

It is said, yet not true, that old time will erase
    From our minds, every thought of a love that is cold,
And the eyes will cease longing to view the fair face
    Of the soul's dearest idol, when love was first told—
But, alas, do we find it in life to be true!
    Does that love grow to hatred or perish with age!
No, never! No matter what course we pursue,
    Like a phantom, 'twill haunt us on life's barren stage.

We may play any part that the fates may assign;
    We may wrestle with fortune or sink low in shame,
Still that love dream of youth, like an angel divine,
    O'ershadows our lives, as the smoke from a flame,
We may go to the West, where the sun never sinks
    In his night robes of darkness, away from our view;
We may fly to the East, yet sad memory links
    Her bloom still to ours, with a chain welded new.

We may drink, we may revel, to drive back the spell;
    We may sleep, yet the dream haunts our being the more,
And the form, and the face, of that one loved so well,
    Stands before us, no matter on sea or on shore.
We may ask of our mothers some cure for such grief,
    As is felt when the heart-strings are severed in twain;
We may seek other friendships to find a relief—
    But alas, to no purpose, all, all, is in vain.

WHEN THE LAST TURBID TIDE.

When the last turbid tide, on the ocean of life,
  Speeds my bark through the straits to Eternity's sea,
May the spirit that guides bear a torch lit with love,
  And whisper, "God bless you, Enola," for me.

And then may the beam of each sentient orb,
  Bathe thy soul with their rays ; light the path for thy feet,
And burn in thy mind, by each radiant gleam,
  One thought of affection for me till we meet.

IS KISSING AN EVIL?

When the dawn of creation was only a dream;
When the first tiny atom began this earth's clod;
When the first ray of light, through the chaos did gleam,
'Twas the kiss from the lips of a wonderful God!
When the earth, from the darkness of chaos, rolled free,
The sun kissed its cheek to verdure and bloom,
And the stars kissed the mountains, and valleys, and sea,
And left light's sweet kisses to burn out the gloom.

And ere there was mortal to sail on the deep,
The moon kissed its slumbering face o'er and o'er;
And the tide queen arose from her pillow of sleep,
To moist with her kisses the surf-beaten shore.
And as the creation unfolded her scroll,
From the kiss of the darkness to life and to light,
The breath of a kiss fanned the flame of a soul,
And he lived as a being of joy and delight.

Then the kiss of deep sleep o'er his lids melted warm,
　　As the flowers are bathed by a summer night's mist,
And a soul from his soul, and a form from his form,
　　Were fashioned and patterned to love and be kissed;
And when he awoke on the dawning of bliss,
　　As a bird waking sings for its mate in the grove,
He greeted fair Eve with a rapturous kiss,
　　And their heart-strings were thrilled with the transport of love.

And love mated each living thing on the earth,
　　By the seal of a kiss—by a kiss it was done;
And the music of Eden grew hushed at the birth
　　Of the gladness of Eve, when she kissed her first son;
And the birds kissed each other and soared in the skies,
　　As she looked back and smiled on the sword o'er the gate,
Content in the light of her infant babe's eyes,
　　And glad of the kiss that had thus sealed her fate!

Then ask me not, maiden, if kissing is wrong.
　　'Tis the fragrance that steals from the heart of the flower;
'Tis the soul's burning current that flows in a song—
　　A song that floats ever through love's lighted bower,
And the woman, whose lips are like cold icy eaves,
　　That drip ever chilly through day's dreamy mist,
Denies the Creator; all nature deceives,
　　And is worthy, alone, by the wind to be kissed?

TWO VESSELS.

It was a summer morn two vessels left the shore,
    Bound for a distant clime far o'er the sea.
The breeze was fair, the sea was calm and o'er
    The world a bright sun glowed resplendantly.
They slowly drift together till at last
    A storm is borne, the mad winds hopeless wail;
The waves surged high and fiercely in the blast,
    The vessels drift, drift onward with the gale.

The sea grew wilder through each passing hour,
    These vessels are divided in the night;
The storm-tossed seamen prayed Almighty power
    To guard and keep them safely in His sight.
That night, my God! That night upon the wave,
    Those hopeful barques that drifted from the shore,
Full soon, alas! had found a drifting grave —
    Their requiem old ocean's sullen roar.

Another day dawned brightly, and the sea
    Grew calm and smooth and peaceful, while the sky
Was as a mirror polished bright might be,
    Cheering the other sailors passing by.
"What shore was it they sailed from, and what sea?"
    Perhaps some anxious friend may ask, "And why
Were they so fated?" It was youth so free,
The sea was life. The vessels you and I.

OH WHERE IS THE TORCH?

Oh where is the torch that once kindled love's fire,
    Since the flame in my bosom has fled?
Where once I adored, I now merely admire,
    And regret every word I have said!
I thought that I loved thee, 'twas surely a jest!
    Thy heart is as cold as a stone!
The fact is too plain, I believe it is best,
    Forever to let thee alone.

Affection of others might cling to a form,
    Though a dull, pulseless statue it be,
But give me a heart that beats constant and warm,
    Whose love has been proven to me,
And as a wrecked sailor clings close to a mast
    Of the vessel that founders at sea,
So my heart will cling ever, and firm to the last
    Dim ray of her love shed for me!

But the glorious angel of love, keeping guard
  O'er the gate of my heart, smiles to hold
The key against those who deserve no reward,
  And are fickle and haughty and cold.
For the deepest of stings that e'er tortured and burned,
  To the quick, true affection's white heart,
Is the touch of the flame of a love unreturned,
  Which burns all its heart strings apart!

OH, TAKE MY HEART.

Oh, take my heart, since it can not be free.
I've tried in vain to hold it back from thee?
Though months have past since first you blest my sight,
You've been the dream of love's own sweet delight,
Unchanged, unaltered, since that far, fair day,
We met and you stole heart and soul away.

Could thought's deep ocean calm herself to rest,
And weep the barks of memory from her breast,
And dash them on Oblivion's rocky coast,
Where hope and love and all that's dear are lost,
Methinks some spirit o'er that nightly sea
Would bring your image hourly back to me.

Or were I cast upon some mystic isle,
Where every flower of heaven loved to smile,
Whose fragrance lulled the sea nymphs to repose,
And angels stopped to kiss each budding rose,
Were you not there, Enola, it would be
A saddened waste; a solitude, to me.

Were I transported to earth's hidden mines,
Where wealth untold in gorgeous splendor shines,
And there make choice between their splendid dyes,
And those I see and worship in your eyes,
I'd say, "O wealth, thy greatness here can fall,
One glance from her, outweighs thy worth in all."

Were I where Venus weaves her robes of light,
From threads of gold and lily leaves of white—
Where music, from his native rock, sets free
His Muse, to call the mermaids from the sea,
Music e'en would sadden; Venus grow less fair,
Unless your smiles were given to me there!

Is this an airy of the poet's mind?
No; it is real. And all such love is blind!
Still as the Peri, struggling against fate,
Pleads with the angel to unbar the gate,
I beg you, keep my heart since it can not be free,
I've tried in vain to hold it back from thee!

WHEN THE HEART'S LAST EMOTION.

When the heart's last emotion of love dies with age,
And the spirit of passion grows cold in my breast;
When the tottering mind can no longer engage
In the musings my soul deems most sacred and best,
Then, then I'll forget thee, nor wish life again
To live o'er the past years of pleasure and pain.

When the last draught of joy has been drained from life's
    bowl,
And sadness about me her mantle shall fling;
When life grows a burden, and grief stings my soul,
And the phantoms of failure around me close cling,
Then love, I will ask thy fond spirit to cheer,
And make life worth living another brief year.

When the thread holding life, asunder shall part,
And all I have cherished drifts into the tomb;
When the curdling life-wine lies still in the heart,
And the eye shut forever in Death's sullen gloom,
Then, not until then, will my thoughts from thee turn,
And Hope's wasting candle at last cease to burn.

A SUMMER EVENING.

The din of the city we left behind,
    As we drove along o'er vale and hill;
Oh, it all comes sweetly back to mind,
    As the ghost of memory haunts me still!
And I muse on the wealth of her sunlit hair,
    And her cheek's red blush, and her lips like wine,
And her brow, like the alabaster fair,
    Seemed lit with a glow that was half divine.

I can see her yet, as on that day,
    The evening came, and the sunset skies
Grew flushed with crimson, then silvery gray,
    As the dreamy shadows began to rise,
And the star queens lifted their veils and smiled,
    Down from their burnished cars above,
And I lost my heart in the tangled wild
    Of her dusky, dreamy eyes of love.

I gathered her fruits from an orchard nook,
    And flowers that bloomed by a shady hedge,
And I'll never forget the smiling look
    She gave as I offered love's simple pledge.
As we passed Oak Hill we could yet decry,
    Through the shadows, the statues of speechless art,
And I little thought as I heard her sigh,
    That her soul was ruled by an icy heart.

DO I LOVE YOU?

Do I love you? Ask the murmurs, pulsing from the heart of night;
Ask the stars, whose bright eyes dazzle, in the mirror of the skies.
Ask the dew drop on the lily, glinting in the morning light;
Ask the perfume, as it rises from a rose's dreamy sighs,
And they'll tell you, truly, Emma, as they are to nature's art,
So art thou to every fibre of my soul's most inner heart.

Do I love you? Ask the brooklet, singing on to kiss the sea;
Or the tide, that spreads her white wings o'er the ocean's thirsty shoal;
Ask the coral isles, that glisten o'er the sea's deep mystery;
And the speechless shell, incasing the pearl's priceless snowy soul,
And they'll tell you, ever, Emma, as they are to nature's art,
So art thou to every fibre of my soul's most inner heart.

Do I love you? Ask the bird's song, floating from the tangled wild;
Ask the notes of sweetest music ever dreamed or played or sung;
Ask the soul of every cadence that e'er thrilled the heart, and smiled;
Or the sweetest, softest curfew that the evening bells e'er rung,
And they'll tell you, truly, Emma, as they are to music's art,
So art thou to every fibre of my soul's most inner heart.

AS WE STROLLED IN OUR SLEEP.

As we strolled in our sleep down Dream-a-way Lane,
   How faithful and fondly, divine of mine,
   You clung to my arm like a tender vine
Clings to the oak in sun and rain,
While Time, grinds over grain on grain,
   The perfumed seed that makes life's wine
   Sweeter than odors of muscadine.
Love, then we heard the soft refrain
   That the night-wren sings when the Muses dine.

And we saw a star from her dizzy height
   Fall from her car in a suicide;
And her life dripped out in drops of light
   That colored the feet of the ocean's tide;
And the moon seemed frightened with her flight,
   As she pulled the clouds o'er her face to hide,
And darker grew the veil of night,
   While the sea-gulls shrieked and the night-owls cried.

Then love, a-quiet about us grew,
    As noiseless as four-o'clocks close their lips,
    When day comes down, and daylight drips
Over the edges of night, like brew
Drips from a goblet; or like the dew,
    Drips from the bud, where the humming-bird sips,
Till tipsy from fragrance it never knew,
    It flies to its mate, and in love's eclipse,
They love to satiety all day through.

And we saw the rarest of flowers in bloom
    Inviting the butterfly and the bee;
But where Sweetness dwelled, there was just room
    For a tiny glance, from you and me,
    As she dozed in a lady-slipper, free
From aught that would cast a shade of gloom,
    Or shorten her life's sweet minstrelsy.

ON THE WORLD.

HE.

Why are you here, fair Magdalene,
  In this detested mart of shame?
Though many others here are seen,
  Why thus debauch an honored name?
Your beauty, and your guileless face,
Look pitiful in such a place.

I read some secret in your eyes,
  That you would from me proudly hide—
A message deep within them lies,
  That tells me Virtue has not died.
I fancy, as I meet your gaze,
Your life has known far happier days.

SHE.

My unknown friend, you reason well.
  With breaking heart, I bear my shame;
It only maddens me to tell
  How once I bore a virtuous name;
Was in my home its pride and light,
But now its sorrow, and its blight.

Sometimes deceptive smiles will play;
  On brow, and lip a light will glow.
To those who stare, I'm blithe and gay,
  While every heart-throb fills with woe.
This I must bear, the fault's my own,
Blame lies with me, and me alone.

Why do I thus to you make known,
  The thing I am or what I've been?
Can you, a man, forgive and own
  That you regret a woman's sin?
No, no, you'll likelier paint my crime,
In darkest tints for future time.

HE.

Fair girl, think not the stain so deep
   That you may not forgiveness win,
For man must make his conscience sleep,
   Ere he e'en names your act of sin.
I leave such work for fools and knaves;
Let sentence come from Him who saves.

SHE.

If thus you think, I'd truly tell,
   The truth to one who'll sympathize;
For I was tempted, and I fell,
   But oh, good sir, you'll not despise
The woman who is thus disgraced,
When you her dreary life have traced.

In early girlhood's happy hours,
   When life and love so truly seemed
A happy path through beds of flowers,
   A lover's eyes upon me beamed.
Him will I cherish through the years;
He gave me love—and shame, and tears.

I met him when my heart was young,
    Unschooled to grief, unknown to care;
Ere it was by love's torture stung,
    And all to me was sweet and fair.
I met him, and my love I gave;
Became his passion's helpless slave.

Full six long years have come and gone,
    Since first I vowed to be his bride.
My life, my love, my all he won,
    And then it was sweet Virtue died.
It was his pleading, from me stole
That priceless treasure of my soul.

I care not what my past may seem,
    I am a slave to love betrayed;
And Love will lead, as in a dream,
    The will into some yielding shade
Where all save love is lost, and yet,
While losing all, we lose regret.

Are any without crime or care,
    From youth until life's race is run?
Such ills and faults, we women share;
    We must have darkness with our sun,
To make our lives a perfect day,
And so we drag the years away.

"Fool's Paradise," it may be called,
    If passion's words and looks are Heaven.
I found it so, and stood appalled,
    When first I learned what love had given.
My father knew it next and he—
He was the first who censured me.

A father, pitied least of all,
    And drove me madly from his gate.
That father, angered at my fall,
    Thus helped me to this crimson fate—
His curse has doomed me to this life,
Who should have been an honored wife.

He fiercely drove me from his door,
    Pronouncing evil on my name:
"Leave me and show your face no more!
    Go from my sight and hide your shame!
No mawkish sentiment I'll waste,
The act is yours, be you disgraced!"

I prayed to God that I might die,
    As outward from my home I went,
For who felt more accursed than I?
    An outcast, hopeless, penitent,
Of love, of home, of friends bereft,
And naught save shame and sorrow left.

I felt my bark of life was wrecked
    On sad Misfortune's jagged reef;
With that bright future we expect,
    Deep darkened in a night of grief,
And I e'er tossed on life's rough wave,
With none to rescue, none to save.

Were mother living that sad day,
   A shelter I would have at home;
A mother could not cast away—
   Her child should not dishonored roam.
Were she alive, life's storms I'd brave;
Her love and sympathy would save.

Ah! she would redirect the act
   A rash impetuous father did.
Her prayers would cause him to retract,
   And steep his eye's most stony lid
In soft'ning tears, that he might see
The wrong to which he's driven me.

My lover? Well, I only know
   He took my heart and went away;
He came no more to me, but oh;
   I do not think he meant to stay.
But when I ponder on his fate,
My heart seems like a leaden weight.

Well I recall that fearful night,
   When Lesley went down in the sea!
The lightnings glared along the height
   Of Heaven's blackened canopy;
The ocean roared, the thunders rolled,
While deep toned bells on life buoys tolled.

Nor can I e'er forget my dream,
   I saw him shivering in the gale;
I saw his wild eyes, frightened, gleam,
   While clinging to a flapping sail;
His night robes loosely 'round him blown,
He feared, yet braved his death unknown.

His brow was pale, and 'round it blew
   His drenched locks, wet with salt sea spray.
I heard him pray that God might do
   Good things for me some future day.
O'erwhelmed at last, beneath the wave,
He sank, and found a sailor's grave.

I woke in fear, with nerves unstrung;
   My mind was like a wilderness,
Dreary, and waste; my bosom wrung
   Still deeper with its sore distress.
The night fled back from morning's beam,
Yet gloom seemed tangled in each gleam.

I never passed such hours of dread;
   Walked to and fro, but all was vain.
The dream to me was riveted—
   Linked to my soul as by a chain.
Anon I saw the stormy coast;
Anon, was haunted by his ghost!

What? You would like to see his child?
   Well, look into his sunny face—
The sweetest face that ever smiled
   Upon this earth with heavenly grace;
And yet, his mother's love and pride
Must cast him, ever, from her side.

Yes, yes, kind sir, it must be done.
   Oh, God! the pangs that tear my breast,
That I must drive away my son,
   And yet I know that it is best.
In time he'll know his father's name,
But must not know his mother's shame.

To every drooping thing on earth
   Is mercy shown save us, alone;
The sparrow, without seeming worth,
   The tenderest loving care is shown;
Each woe and care a tear will claim,
Except where woman soils her name.

Yes, but for honor's sense alone,
   I would be loathe from him to part.
The human breast can not be stone,
   For if it could it would not smart;
For honor's sake our hearts must cry,
And feel remorse until we die.

And—pardon me, sir, I can see
　The same soft love-light in your eyes
I saw in his.　Speak, can it be,
　Or is my poor heart breathing lies?
Yes! yes! thank God! it's gladly true—
The sea gives back my love—and you.

HE.

True, sweet Loretta, I am he,
　Who made your true heart bleed and break.
No matter what the end may be,
　I'll wed you now for love's sweet sake.
The fault was mine, by my neglect
I was your sad fate's architect.

LABOR.

THE VOICE OF LABOR.

From the mine and from the factory, from the field the cry is heard:
"On to freedom, on to justice, on and on!" We hear the word,
Rolling, like the pulsing current of a never ending sea.
On to justice, on to freedom, man shall have his liberty!
And the voice of right, once echoed, never can be stilled to rest,
'Till the social ban is lifted and the toiler's wrongs redressed.

No longer men are silent, trudging on in mute despair,
Making no complaint or murmur, of the galling yoke they wear.
But the voice of protest deepens, as the cycling ages roll,
And their liturgies of sorrow dirge-like sway the saddened soul.
As the tide that sweeps the bosom of the deep and darksome sea,
So shall Justice sweep Oppression till humanity is free!

Brave minstrel of a free-born race, thy harp is now unstrung.
Else why do millions live and die unhonored and unsung?
The threads which hold existence, seem to part as by a flame,
And the shades that pall their death-couch, are of poverty and shame!
But thy harp's sweet chords will vibrate once more, in our lovely land,
When thy form's unbent, uncrippled, and less calloused is thy hand.

Melodies, as sweet to freedom as the strains from Heaven's choir,
Flowing from the heart of feeling, kindled by the soul's own fire;
Melodies, to live forever, echoed by the air and light,
And enchant the drowsy moonbeams, as they wander through the night;
Melodies, to rock the forests and make drunk the soul of dawn,
As when the walls of Thebes rose to smile on Amphion!

Shame! Oh, shame to this great people that they do not know their power,
But the horologe of ages soon will register the hour
When the dawn of truth and reason will dispel the shades, that blind
The stream of thought, that rushes from the chaos of the mind.
Then, with bronzed and haggard visage, those who've borne their task sublime,
In the battle march of progress, o'er the battlements of time,
Will look back and sigh with pity at the skeleton of Wrong
That once drove and bound a people with his lash and golden thong.

Right and Justice then will triumph, and the people will not grope
In the labyrinth of turmoil, stripped of all in life save hope,
And the weary surf-worn exile, driven from his native shore,
Will not in Columbia's vineyard meet the tyrant at his door.
For this goodly land will slumber, wrapped in Nature's choicest law,
And each heart, congealed to feeling, in the light of reason thaw.

Then man's social rank and standard will not be the dollar's stamp,
Nor will labor's stencil brand him as an abject slave or tramp;
And the blood that swells his life-tide will not be the Britton's blue,
But America's untainted, loyal, brave, and warm and true.

Oh, God! turn thy face in sorrow, witness not the laborer spurned
From the doors of gilded mansions his own calloused hands have earned —
He, who has by years of struggle laid the giant forests waste,
Garnishing the hills with cereals and with fruits of sweetest taste;
Who has builded splendid cities and has wrested from the hills,
Precious metals, now corroding, unused in the banker's tills!

Friendship, Oh, thou white winged angel! Wilt thou hover o'er our lands,
Place two souls in one strong body, and unite the toilers' hands?
Let them read the blood-stained parchment, that was written in the past,
When the manacles of slavery bound and held the negro fast!

Then, as men endowed with reason, close the pages they
 have read,
And revere our noble heroes who now slumber with the dead;
And on this great nation's altar take an oath, that never more
Shall our history's page be blotted by war's hands and
 streams of gore!
Dare we say, in this grand issue, that we have a North and
 South?
No! From the great lakes of th' Northland, to the Mississippi's mouth,
Every toiler feels the iron heel of monopoly and wrong,
And their spirits blend together in the shout of freedom's
 song.
O'er the chasm is suspended brotherhood's great bridge of
 right,
And its girder's are of muscle and its piers are hearts of
 might!
And beneath it lie the cannon, and the muffled drum at rest,
That have made the earth to tremble, thrilling every warrior's
 breast.
Let us leave them there, brave toilers, slumbering ever in
 the mould,
Wreathed alone by saddened memories of a cruelty untold.

AN APPEAL TO PATRIOTISM.

Back through the mist of sorrow-laden years,
Which sleep in darkness, stained by blood and tears,
My vision lingers o'er the ruins vast—
The once proud glory of the ages past;
And mourns the loss, because Oppression's hand
Strove to enslave the free and take their land.

Do we forget that we were born the same
As our forefathers, who, by dint and name,
Proved to the world by their New England birth,
That they were men of valor and of worth;
And that no despot e'er should rule a son
At Bunker Hill, Concord and Lexington?

As men of learning and by wisdom taught,
Shall we forget our land so dearly bought?
A land, which while in infancy, uprose
The power of tyrants sternly to oppose,
Believing justice should be given all,
As that declared at Independence Hall.

For Freedom's cause did not a Hancock grace
And Adams prove a Father to his race?
Did not a Henry snatch the tyrant's veil
To see his iron features twitch and pale,
And Jefferson, with giant mind make plain
The cause of Justice free from blot or stain.

Are we to love the country of our birth,
Robbed of our prided heritage to earth,
Or love a land where lords proscribe and rule,
Or laugh man's sacred rights to ridicule,
Allowing class to legislate, secure,
To class alone and leave the millions poor?

Think of the isle where Roger Williams knelt
To pray for mercy and for justice felt
Within the hearts of men, and call
For life and liberty to one and all.
May that proud island sink within the sea,
Than have it said her men are not all free!

Gaze on the scaffold, where the martyr smiled
To give his life for freedom, while the wild,
Untrammeled sunlight bore his soul away
Into eternal rest. Oh, glorious day!
That dawned o'er Harper's Ferry, and unfurled
The scroll of John Brown's life to all the world.

Are we to love the flag that Fremont bore
O'er desert sands to reach the Western shore?
The same that Jackson, pride of North and South,
Waved o'er the foe at Mississippi's mouth?
Yes, that dear flag is still our flag to-day.
Once loved by Webster, Lafayette and Clay.

Yet let its folds, from out oppression's night,
Unfurl the stars of justice and of right;
Let every toiler on the land and sea
Clasp hands to keep this emblem of the free;
Nor let its folds be trailed within the dust,
While freedom triumphs and "In God we trust."

We scan the history and learn this truth—
That all the wars since first the country's youth
Have been brought on by stern oppression's curse,
Slavery and wrong, grim violence and force,
Until the people, desperate for right
Rose up to triumph in their pride of might.

My fellow brother, which shall be the end—
The sword? Or shall intelligence befriend?
Shall we resort to war's destructive plan,
When honest ballots will the right to man?
Shall reason argue with each human mind,
Or shall we live and die as serviles blind?

Oh, glorious land, the birth-place of the free,
Long hidden from the gaze of tyranny!
Must I resign my freedom at thy door,
With millions flocking here from every shore,
Who come serf-scourged, and still as serfs must slave,
And give all else, save hope beyond the grave?

Shall this Republic, with her wealth untold,
Fall as the Roman Empire did of old,
Because patrician glory stood so high
And plebians were doomed to do or die?
Or shall we check the moneyed Neros now
And ask for rights to him who holds the plough?

Review again our history's darkened page.
Within the past, it has not been an age
Since Sumpter woke, by dread and sullen roar,
Each patriot son who loved his native shore,
And war's red hand smote mountain, field and glen,
Destroying peace and millions of our men.

Awake, Oh patriot spirits from the gloom!
And breathe a flame as breathes the hot simoom,
That will destroy each flagrant, selfish aim
In all the earth; and may the tyrant tame
His fiendish thirst for life and land and gold;
And mankind dwell at peace in nature's fold.

THE WANDERER.

Give me drink! No, do not curse me—do not ask if it be right,
But obey my prayer, barkeeper, let me quench my thirst to-night!
Let me quell that thirst which maddens, yes, which fevers heart and brain,
Grant it, Oh, for God's sake grant it! Let it drown this sting of pain.
Just one goblet, give it to me, and though serpents in it dwell,
It were sweet, and I will quaff it, though they bind my soul in hell,
For its fiery kiss is sweeter than the fountain streams of youth,
And its lying lips more welcome than the lips of love and truth.
I'm a tramp? Well, yes, a tramp, sir, but don't let that stay your hand,
For there are five hundred thousand like myself in this fair land!
And you must remember one thing—evil springs alone from cause,
And that cause is quite apparent in our social statute laws.

Circumstances make the standard we should measure all
    men by,
And society will lower, or will elevate most high.
Man's a creature of conditions and environment. 'Tis true
I'm a tramp, but not by option—labor is what I'd pursue.
May God bless you for this drink, sir, it's an antidote for ills ;
It will drive back that Nemesis that now burns the soul and
    kills ;
It will curb the recollections of my sad eventful years,
And will dry my eyes from weeping disappointment's bitter
    tears.
Think, sir, what I've had to combat, since a lad about half
    grown.
Father lost his life in battle leaving ma and me alone :
And from that day on to this time—and it may be to the
    grave—
I have known, and no doubt will know nothing else than be
    a slave.
Not so much the slave of drudging, as a slave to blasted hopes,
By that cursed fate which follows and forever near me gropes.

We were dwelling in a city beautified with domes and spires,
Where ten thousand happy families peacefully enjoyed their fires;
And throughout that lovely valley, peace and fortune seemed to reign,
Until high-wrought desolation turned loose his destructive train.
Fact is, that some thoughtless people, for their own sweet pleasure's sake,
Dammed a stream, far up the valley, turning it into a lake,
Where its giant waters slumbered, as an evil force asleep;
Watched and waited like a tiger, crouching for his deadly leap.
Hark! Great God! I hear those waters as they madly rushed that day
Down the Conamaugh river valley, sweeping Johnstown quite away;
And I see that fearless horseman, and I hear his warning cry,
As destruction sped behind him: "Flee for refuge! Flee or die!"

And I see those steepled buildings, reeling, toppling to and fro ;
And I hear them clash together as the knights of long ago ;
And I hear the shrieks of women, and I hear the prayers of men.
But there was no hand to rescue from the watery, deathly den !
Give me drink ! Oh give me liquor ! for I can not bear the thought ;
Let me blast the recollection of the havoc it has wrought !
For on that day in my cottage dwelt the ones I loved the best—
She, the angel of my life, sir, with a young babe at her breast,
And my old and feeble mother—Oh, my God, how could it be
That the hand of fate decreed, sir, that they should be swept from me ?

Do you blame me then for drinking ? There's a fate for every man,
And to blast my joys and pleasures seems a part of Satan's plan.
But this is not all the story. After that eventful day,
With my hopes and prospects wasted, do you think I there could stay,
Where the ghost of that fair angel, and the babe upon her breast,
Would come back to me each moment, till I found no peace and rest ?

No! So then I went to Homestead, where a man his manhood breaks,
And declares himself a servile for the pittance that he takes.
You have read about the strikes there; you have learned the facts, no doubt,
Of how their rights were trampled on, and the working men locked out.
And the Pinkertons were hired, like dogs to shoot them down,
Until Homestead seemed a Hell, sir, by the riot in the town.
Of those toilers I was one, sir, who was locked out in the strike,
Which now makes me tramp the country, something that I do not like.
But I'd rather tramp a free man, than to toil a rich man's slave,
And forever kiss a lord's foot, and then fill a pauper's grave!

IN CHAINS.

I had a dream, a troubled dream, that made my bosom smart—
A vision, weighted down by chains, came in and touched my heart;
My blood seemed cold, my lips grew dumb, my eyes could only stare,
And leer enchanted in the light of her eyes and golden hair.
The chains were not of that dusky hue, that clanks at the felon's feet,
Or galls the wrists of the souls depraved, that we see along the street,
But heavy and bright they glowed, with light like the fabled fleece of old.
Were they moulded, and folded, and welded there—those binding chains of gold?

At length my lips, no longer dumb, grew bold enough to
   speak:
"Who art thou, child of night or day? And what is it ye
   seek?
What spectre comes at this late hour to trouble thus my sleep,
And what can be thy mission from the dark lethean deep?
Hast thou escaped from Heaven's gate to seek release on
   earth,
Or have the fiends of Hades chained thy form with all this
   worth?
Art thou Aladdin's ghost disguised, some secret to impart?
Speak, spirit, speak, and let the blood once more warm up
   my heart!"

" Fear not," the apparition said, " you know me not I see,
But I'm the goddess of your land, the Goddess Liberty.
I am no cad of Darkness, but a messenger of Light,
And it is sad that I now wear this livery of night.
The great red Dragon, known as Gold, has bound me thus
   in chains,
And smirched my whitened robes of power in black, dis-
   graceful stains.
The wreath that angels gave me to wear upon my brow,
This Dragon imp has snatched away and it is wasted now.

"Greed has no care for Freedom's flowers, but with a prided lust,
With iron heel he tramples them into the mire and dust.
Behold my empty calloused hand that jewels once impearled
And held the torch of Liberty enlightening the world.
My star-infusing hope and joy are hid in clouds of shame,
That I no longer guide earth's hosts to glorious deeds of fame
But Oh, brave Knight, it is God's will that I should share this fate
A season only, as I watch by Truth's eternal gate.

"No power can take from me the scale by Love and Justice given,
In which I weigh man's rights and wrongs now registered in Heaven.
Full many a saber scar's been hacked upon their golden beams,
As I weighed out the blood of wars and great reformer's dreams.
The tears and cries of hopeless men, and anguish manifold,
Are borne against the oppressor's wand, and heaps of rusting gold.
To God's great time-lock, I alone possess the magic key,
And when the hour arrives, then will I set all nations free.

"The treasures hid in Mammon's vault shall not enslave the land
So long as I, a goddess reign, with this key in my hand.
The cursed alchemists too long have ruled this goodly earth,
Decreeing social rank and file by high and lowly birth.
At gold's polluted shrine they've stood pretending Christian grace,
To mock and spit upon a Christ who died to save His race.
They've clad their forms in purple and imbibed their wine at ease,
While men, to live, have knelt to them and slaved on bended knees!

"But soon the segment of their power will fall, and my elect
Who have created all the wealth will rise and stand erect.
As God designed that they should stand as from His hand they came,
Ere lord or knave or factory boss laid on them lash or claim.
On Time's prophetic dial I see the shadow moving on,
And soon his clock with waking din shall strike the gladdening dawn;
Then will my star of freedom shine on once more, like the sun
Shone o'er the hosts that fought and bled with Stark and Washington.

The standard bearer of the age with Freedom's flag unfurled
Will bring back on the wings of flame my torch to light the world.
And that red Dragon born of Hell, in these cursed chains shall die,
And I may prove Bartholdi's gift no monumental lie.
The lily and the olive wreath again I'll wear with pride,
And on Columbia's shore no more my rights will be denied.
The British Lion, for his prey, shall tread our shores no more,
Nor eagles, into vultures turned, guard by the rich man's door.

No more shall chartered crime prevail against the rights of men,
Nor babes, of sorrowing mothers, die in Famine's horrid den.
Hope's candles then shall not burn out in Disappointment's gloom,
And leave a race to grope unheard within a living tomb.
The slave Prometheus shall rise, freed from his chains, and slay
The vulture Greed that's feasted on his liver day by day!

LABOR DAY.

From East to West, from North to South;
   O'er rivers, lakes and seas,
Let shouts ring out from mouth to mouth,
   Fling banners to the breeze!
Let factory slaves again be men,
   And show the world the way
To liberate itself again—
   Turn out for Labor Day!

Brave artisans from shop and soil,
   Lay all your cares aside;
Come swell the ranks of honest toil,
   And view your works with pride.
The palaces of brick and stone,
   And all of structural worth,
Your calloused hands, and yours alone,
   Have builded on the earth.

The farmer comes from furrowed field;
    The smith from forge of fire;
The miner's face to light revealed,
    Comes lit with stern desire;
The cooper from his hoop and stave;
    The potter from his clay—
These men were never born to slave
    On this great Labor Day!

They've seen the products of their hands
    Heaped into Plenty's lap;
They've slept while vampires of the lands
    Did their rich life-blood sap.
But chainless, they awaken now,
    To grasp their flag unfurled,
And like true men, no longer bow
    To despots of the world!

Throw down your tools like men of might,
    And let your worth be known,
And show those lords you have the right
    To reap what you have sown;
That God created all men free,
    A truth none dare gainsay—
Hallowed by Him from sea to sea
    To celebrate this day!

What if the autocrat should sneer
    At honest labor's hand?
He knows that should it rest a year
    'Twould bankrupt all the land!
Aye, rest ten days from all its toil,
    What would the mocker's see?
No cars would speed across the soil,
    No ships would sail the sea!

No messenger from foreign shores
    Would wear an inky dress;
No factories stand with open doors,
    And hushed would be the press.
No royal ladies of the day,
    Would doff their silks and pearls,
And at the wash-tub kneel and pray
    Once more for servant girls.

"God save the King"—His grace would come
    From his high throne to greet
His equal—a poor toiler dumb—
    And kiss his calloused feet!
The glare of gold would fade away,
    And men would own the might
That rests with toil on Labor Day—
    The awful might of right!

www.ingramcontent.com/pod-product-compliance
Lightning Source LLC
Chambersburg PA
CBHW031454160426
43195CB00010BB/975